Energy:
A Bibliography of Social Science and Related Literature

Garland Reference Library of Social Science (Vol. 9)

Edited and Compiled by Denton E. Morrison:

THE SIGNIFICANCE TEST CONTROVERSY: A READER

(Ramon E. Henkel, Co-Editor)

ENVIRONMENT: A BIBLIOGRAPHY OF
SOCIAL SCIENCE AND RELATED LITERATURE

(Kenneth E. Hornback and W. Keith Warner, Co-Compilers)

Energy:

A Bibliography of Social Science and Related Literature

Denton E. Morrison

With the Assistance of
Virginia Bemis, Olivia Mejorado,
Sharon Loomis and Odessa Bogan

Department of Sociology
Michigan State University
East Lansing, Michigan

Garland Publishing, Inc., New York & London

1975

Copyright © 1975 by

DENTON E. MORRISON

All Rights Reserved. No part of this publication
may be reproduced or transmitted in any form
by any means, electronic or mechanical, including
photocopy, recording, or any information storage
and retrieval system, without permission in writing
from the copyright holder and publisher.

Library of Congress Cataloging in Publication Data

Morrison, Denton E
   Energy, a bibliography of social science and related
literature.

   Includes indexes.
   1. Power resources--Bibliography. 2. Energy policy--
Bibliography. I. Title.
Z5853.P83M62   016.3337   75-4800
ISBN 0-8240-1096-5

Printed in the United States of America

To

W.F. COTTRELL

ENERGY PIONEER IN SOCIOLOGY

CONTENTS

	Page
Preface	ix
Bibliography	1
Addendum	129
Subject Index Categories	158
Subject Index	161

PREFACE

There is an overwhelming, unfortunate tendency to think of energy problems and solutions in physical science and engineering terms. It is increasingly apparent, however, that energy problems are in substantial measure the consequences and the causes of social phenomena: social attitudes, social behaviors, social institutions, social structures and populations. To address energy problems outside the context of social factors is to guarantee superficiality in the level of understanding achieved and failure in the solutions attempted.

One of the purposes of this bibliography is to draw attention to the fact that we do not start from zero in social science understanding of energy problems. While we need to know more, much more, about the social science aspects of energy, a good part of what we need is not new information that has to be discovered by new research. Much of what we need is to become better and more widely informed about what is already known. This bibliography clearly indicates that the current involvement of social scientists is high, but it also shows that many social scientists have long been concerned with energy problems and have produced a considerable body of knowledge that pre-dates the current energy crisis. Part of the energy problem, then, is that we have not made widely known or paid heed to what some have known for some time -- and this is true of both the physical and social science aspects of energy.

We hope, then, that much of the literature in this bibliography will be immediately relevant to current energy problems. But, aside from such direct applications, is the fact that current and future research efforts cannot possibly have maximum scientific or practical efficiency unless these efforts stand on the shoulders of the accumulated knowledge. Research that does not build on past knowledge represents a great waste. We hope that by this compilation we have helped reduce such waste.

No bibliography in a field as broad and as burgeoning as social science and energy can be either exhaustive or current. The scope of the bibliography is, however, deliberately comprehensive across the full range of English language social science and related literature. This is because it is neither possible nor fruitful to put firm boundaries on "social science" or "energy". In addition to entries that link social with physical energy information, some entries report physical energy information of the fundamental sort that must be taken into account in various social science analyses. The "Subject Index Categories" list on pages 158 - 160 provides a concrete guide to the scope of the bibliography.

We have attempted to be as up-to-date as possible by including an "Addendum" for items that came out or came to our attention after the original compilation in the fall of 1974 and prior to completion of the manuscript in January of 1975. Certainly, however, an additional attempt equal in time and effort to that which produced this bibliography would produce one much larger and more complete. But we have taken the view that both the pressing, current reality of the energy crisis and the need for maximum, intelligent social science involvement in addressing it make it important for social scientists to have a reasonably complete bibliographic resource at the earliest point in time. Consequently, we hope that those who are frustrated by the inevitable errors, incomplete entries, omissions and irrelevancies will direct their aggression in a creative channel by sending to us for the next edition the improved information that will assumedly emerge from their efforts in dealing with our efforts.

We received help from many sources. In particular several other more specialized or more general energy bibliographies were important. Bibliographies by Norman Dalsted and Larry Leistritz and by Nan Burg were especially valuable and we thank their compilers for making them available with permission for use. We also wish to acknowledge the utility of bibliographies compiled by Dana Ellingen and Willian Towsey. Many individuals and organizations sent lists and papers. We are especially grateful to Paul Bugg, J.P. Forgarty, Sam Klausner, Richard Kouri, Dean Mann, Frank Osterhoudt, Bob Otte and R.P. Ouellette and to Environment Canada. Finally, we are grateful for the continuing support of the Michigan Agricultural Experiment Station.

BIBLIOGRAPHY

BIBLIOGRAPHY

1. Aaronson, Terri. "Tempest Over a Teapot." ENVIRONMENT 11: 22-27. October 1969.

2. Abel, Fred. "Contemporary Directions in Policy Making Affecting Agriculture: Managing Environmental Resources." Paper Presented to the North Central Regional Research Strategy-2 Committee on Natural Resource Development, Zion, Illinois. April 1973.

3. Abrahams, P.P. "Balancing Environment and Energy Needs," CURRENT 150: 58-60. April 1973.

4. Abrahamson, Dean E. ENVIRONMENTAL COST OF ELECTRIC POWER. Scientists' Institute for Public Information, New York. 1970.

5. Abrahamson, Dean E. "Overview: Energy." ENVIRONMENT: January, March, June 1973. (Three Parts).

6. Abrahamson, Dean E. "The Environment: When is Doomsday?" In ENERGY AND THE ENVIRONMENT: A COLLISION OF CRISES, Irwin Goodwin (ed.). Publishing Sciences Group, Acton, Massachusetts. 1973.

7. Adams, R.G. "Developments in Electric Power Generation in Ontario." GEOGRAPHY 57(256): 235-239. July 1972.

8. Adams, Richard N. "Energy, Culture and Social Power." Paper Presented at the Annual Meeting of the American Association for the Advancement of Science, San Francisco. February 1974.

9. Adamson, Colin. "Electric Energy Trends in the U.K." POWER ENGINEERING 75: 26-32. February 1971.

10. Adelman, Morris Albert. SECURITY OF EASTERN HEMISPHERE FUEL SUPPLY. Department of Economics, Massachusetts Institute of Technology. Cambridge. 1967.

11. Adelman, Morris Albert. "Is the Oil Shortage Real? Oil Companies as OPEC Tax-collectors." FOREIGN POLICY 9: 69-107. Winter 1972-73.

12. Adelman, Morris Albert. THE WORLD PETROLEUM MARKET. Published for Resources for the Future by the Johns Hopkins Press, Baltimore. 1972.

13. Adelman, Morris Albert. "American Import Policy and the World Oil Market." ENERGY POLICY 1(2): 91-99. September 1973.

14. Adelman, Morris Albert. "Cartels and the Threat of Monopoly." In ENERGY AND THE ENVIRONMENT: A COLLISION OF CRISES, Irwin Goodwin (ed.). Publishing Sciences Group, Acton, Massachusetts. 1973.

15. Adelman, Morris Albert. "Politics, Economics and World Oil." THE AMERICAN ECONOMIC REVIEW 64(2): 58-67. May 1974.

16. Akins, James E. "Can We Depend on 'Cheap Foreign Oil'?" THE CONFERENCE BOARD RECORD 9(7): 23-24. July 1972.

17. Akins, James E. "The Oil Crisis: This Time the Wolf is Here." FOREIGN AFFAIRS 51(3): 462-490. April 1973.

18. Akins, James E. "International Cooperative Efforts in Energy Supply." ANNALS OF THE AMERICAN ACADEMY OF POLITICAL AND SOCIAL SCIENCE 410: 75-85. November 1973.

19. Alaska Science Conference, 20th, University of Alaska, 1969. CHANGE IN ALASKA: PEOPLE, PETROLEUM, AND POLITICS. University of Alaska Press, College, Alaska. 1970.

20. Albrecht, Stan L. "Environmental Issues: Power Plant Development in the Southwest." Paper Presented at the Annual Meeting of the Pacific Sociological Association, Portland, Oregon. April 1972.

21. Albrecht, Stan L. "Sociological Aspects of Power Plant Siting." Department of Sociology, Utah State University, Logan. 1972.

22. Albrecht, Stan L. "Environmental Social Movements and Counter Movements: An Overview and Illustration." JOURNAL OF VOLUNTARY ACTION RESEARCH 2-11. October 1972. Reprinted in Robert Evans (ed.). SOCIAL MOVEMENTS: A READER, Rand McNally, Chicago. 1973.

23. Alderson, V. Ray. "Adaption of the Motor Carrier Industry to Present and Prospective Supplies of Fuel." TRANSPORTATION JOURNAL 13(3): 20-23. Spring 1974.

24. Alessio, Frank J. "Is There an Energy Crisis Ahead: For the Environment?" ARIZONA REVIEW: April 1972.

25. Allen, Bruce T. and Melnik, Arie. "Economics of the Power Reactor Industry." QUARTERLY REVIEW OF ECONOMICS AND BUSINESS 10: 69-79. Autumn 1970.

26. Allen, Donald R. "Environmental Policy Act Affects Local Public Systems." PUBLIC POWER 28: 24-27. November 1970.

27. Allen, Howard P. "Electric Utilities: Can They Meet Future Power Needs?" ANNALS OF THE AMERICAN ACADEMY OF POLITICAL AND SOCIAL SCIENCE 410: 86-96. November 1973.

28. Altman, Manfred, Telkes, Maria and Wolf, Martin. THE ENERGY RESOURCES AND ELECTRIC POWER SITUATION IN THE UNITED STATES. University of Pennsylvania Press, Philadelphia. 1971.

29. American Association for the Advancement of Science. Committee on Environmental Alteration. POWER GENERATION AND ENVIRONMENTAL CHANGE: SYMPOSIUM. M.I.T. Press, Cambridge, Massachusetts. 1971.

30. American Enterprise Institute for Public Policy Research. UNITED STATES INTERESTS IN THE MIDDLE EAST. American Enterprise Institute for Public Policy Research, Washington, D.C. 1968.

31. THE AMERICAN FEDERATIONIST. "Energy Crisis: The AFL-CIO Program." THE AMERICAN FEDERATIONIST 81(3): 1-2. March 1974.

32. American Gas Association. THE NATURAL GAS SUPPLY PROBLEM. American Gas Association, New York. 1971.

33. American Gas Association et al. TOWARD RESPONSIBLE ENERGY POLICIES: A JOINT ENERGY POLICY STATEMENT. American Gas Association, Arlington, Virginia. n.d.

34. American Mining Congress. THE ENERGY CRISIS. American Mining Congress, Washington, D.C. 1972.

35. American Nuclear Society. NUCLEAR POWER AND THE ENVIRONMENT: QUESTIONS AND ANSWERS. American Nuclear Society, Hinsdale, Illinois. 1973.

36. American Nuclear Society. PROCEEDINGS OF THE INTERNATIONAL CONFERENCE ON NUCLEAR SOLUTIONS TO WORLD ENERGY PROBLEMS, Washington, D.C. 1972. American Nuclear Society, Hinsdale, Illinois. 1973.

37. American Petroleum Institute. INCOME TAX POLICY FOR THE PETROLEUM INDUSTRY DURING THE 1970'S. American Petroleum Institute, Washington, D.C. n.d.

38. American Petroleum Institute. PETROLEUM FACTS AND FIGURES. Annual.

39. American Petroleum Institute. STATEMENT OF POLICY ON PERCENTAGE DEPLETION. American Petroleum Institute, Washington, D.C. 1961.

40. American Petroleum Institute. STATEMENT OF POLICY: CONSERVATION, DEVELOPMENT AND PRODUCTION PRACTICES. American Petroleum Institute, Washington, D.C. 1963.

41. American Petroleum Institute. STATEMENT OF POLICY: PETROLEUM IMPORTS. American Petroleum Institute, Washington, D.C. 1964.

42. American Petroleum Institute. STATEMENT OF POLICY: CONSERVATION OF NATURAL RESOURCES AND WILDLIFE. American Petroleum Institute, Washington, D.C. 1965.

43. American Petroleum Institute. STATEMENT OF POLICY: GOVERNMENT-SPONSORED RESEARCH. American Petroleum Institute, Washington, D.C. 1967.

44. American Petroleum Institute. STATEMENT OF POLICY ON JURISDICTION OVER THE NATURAL RESOURCES OF THE OCEAN FLOOR. American Petroleum Institute, Washington, D.C. 1969.

45. American Petroleum Institute. THE ENERGY SUPPLY PROBLEM. American Petroleum Institute, Washington, D.C. 1970.

46. American Petroleum Institute. UNITED STATES PROVED RESERVES OF CRUDE OIL, NATURAL GAS AND NATURAL GAS LIQUIDS. American Petroleum Institute, Washington, D.C. 1970.

47. American Petroleum Institute. ANNUAL STATISTICAL REVIEW -- U.S. PETROLEUM INDUSTRY STATISTICS 1956-1970. American Petroleum Institute, Washington, D.C. 1971.

48. American Petroleum Institute. Committee on Public Affairs. BIBLIOGRAPHY OF ENERGY AND FUELS. American Petroleum Institute, New York. 1961.

49. American Petroleum Institute. Committee on Public Affairs. OIL AND THE CHALLENGE OF ALASKA: THE PETROLEUM INDUSTRY'S CONCERN FOR THE ENVIRONMENT. American Petroleum Institute, Washington, D.C. 1971.

50. American Public Power Association. STATEMENT OF NATIONAL POWER POLICY. American Public Power Association, Washington, D.C. 1970.

51. American Public Power Association and National Rural Electric Cooperative Association. ARTIFICIAL RESTRAINTS ON BASIC ENERGY SOURCES. American Public Power Association, Washington, D.C. 1971.

52. Amuzegar, Jahangir. "The Oil Story: Facts, Fiction and Fair Play." FOREIGN AFFAIRS 51(4): 676-689. July 1973.

53. Anderson, Dennis. "Models for Determining Least-Cost Investments in Electrical Supply." BELL JOURNAL OF ECONOMICS AND MANAGEMENT SCIENCE 3: 267-299. 1972.

54. Anderson, Kent P. "Some Implications of Policies to Slow the Growth of Electricity Demand in California." Rand Corporation, December 1972.

55. Anderson, Kent P. "Residential Demand for Electricity: Econometric Estimates for California and the United States." THE JOURNAL OF BUSINESS 46(4): 526-553. October 1973.

56. Anderson, R., Brown, J., Tankersley, G. and Waddington, C. "Energy for Technological Development." EKISTICS 34(203): 240. October 1972.

57. Anderson, Richard H. "The Promise of Unconventional Energy Sources." BATTELLE RESEARCH OUTLOOK 4(1): 1972.

58. Anderson, Robert O. "Energy and Environment Need Not Be In Conflict." CATALYST FOR ENVIRONMENTAL QUALITY 3(2): 1973.

59. Andrews, Richard N.L. ENVIRONMENTAL POLICY AND ADMINISTRATIVE CHANGE: THE NATIONAL ENVIRONMENTAL POLICY ACT OF 1969, 1970-71. Unpublished Ph.D. dissertation. Department of City and Regional Planning, University of North Carolina, Chapel Hill. 1972.

60. Angelini, Arnaldo M. "Electricity Generation and Distribution in Italy." REVIEW OF THE ECONOMIC CONDITIONS IN ITALY 28(1): 14-27. January 1974.

61. ANNALS OF THE AMERICAN ACADEMY OF POLITICAL AND SOCIAL SCIENCE. "The Energy Crisis: Reality or Myth." ANNALS OF THE AMERICAN ACADEMY OF POLITICAL AND SOCIAL SCIENCE 410: November 1973. (Whole Issue).

62. Anthony, Robert H. ENERGY DEMAND STUDIES - AN ANALYSIS AND APPRAISAL. U.S. Government Printing Office, Washington, D.C. 1972.

63. APPALACHIA. "The Demand for Energy and Appalachia's Coal." APPALACHIA: February/March 1972. (Whole Issue).

64. Archer, Mary. "Alternative Futures: Prospects for Solar Energy." FUTURES 6(3): 261-266. June 1974.

65. ARCHITECTURAL FORUM. "Architecture and Energy." ARCHITECTURAL FORUM: July/August 1973. (Whole Issue).

66. Areskoug, Kaj. "U.S. Oil Import Quotas and National Income." SOUTHERN ECONOMIC JOURNAL 37: 307-317. January 1971.

67. Armstead, H.C.H. "Geothermal Heat Costs." ENERGY INTERNATIONAL: February 1969.

68. Armstead, H.C.H. (ed.). GEOTHERMAL ENERGY. UNESCO, Paris. 1973.

69. Arnaoot, Ghassan. "The Organization of Petroleum-Exporting Countries (OPEC). THE ROCKY MOUNTAIN SOCIAL SCIENCE JOURNAL 11(2): 11-18. April 1974.

70. Arnold, Dean E. "Thermal Pollution, Nuclear Power, and the Great Lakes." LIMNOS 3: 3-12. Spring 1970.

71. Artsimovich, L.A. "Controlled Nuclear Fusion: Energy for the Distant Future." BULLETIN OF THE ATOMIC SCIENTISTS 26: 47-55. June 1970.

72. Aspin, Les. "A Solution to the Energy Crisis: The Case for Increased Competition." ANNALS OF THE AMERICAN ACADEMY OF POLITICAL AND SOCIAL SCIENCE 410: 154-168. November 1973.

73. ASSOCIATION MANAGEMENT. "How Associations Are Working to Solve the Energy Crisis." ASSOCIATION MANAGEMENT 25(5): 79-85. May 1973.

74. ASSOCIATION MANAGEMENT. "Recommendations for Sound Energy Policies." ASSOCIATION MANAGEMENT 25(5): 86-90. May 1973.

75. Association of the Bar of the City of New York. Special Committee on Electric Power and the Environment. ELECTRICITY AND THE ENVIRONMENT: THE REFORM OF LEGAL INSTITUTIONS: REPORT. West Publishing Co., St. Paul, Minnesota, 1972.

76. Atomic Industrial Forum. ELECTRIC POWER AND THERMAL DISCHARGES. Gordon and Breach, New York. 1969.

77. Auer, P.L. "An Integrated National Energy Research and Development Program." SCIENCE 184(4134): 295-301. April 19, 1974.

78. Averitt, Paul. COAL RESOURCES OF THE UNITED STATES. U.S. Geological Survey Bulletin Number 1275. U.S. Government Printing Office, Washington, D.C. January 1967.

79. Averitt, Paul and Carter, M. Devereux. SELECTED SOURCES OF INFORMATION ON UNITED STATES AND WORLD ENERGY RESOURCES: AN ANNOTATED BIBLIOGRAPHY. U.S. Geological Survey Circular Number 641. U.S. Government Printing Office, Washington, D.C. 1970.

80. Aymond, A.H. "Electric Energy and the Environment." PUBLIC UTILITIES FORTNIGHTLY 85: 39-43. June 4, 1970.

81. Ayres, E. and Scarlott, C.A. ENERGY SOURCES: THE WEALTH OF THE WORLD. McGraw-Hill, New York. 1952.

82. Bachir, Bahij. "Arab Oil." REVIEW OF INTERNATIONAL AFFAIRS 24(565): 25-26. October 20, 1973.

83. Bachman, W.A. "The War on Pollution." OIL AND GAS JOURNAL 68: 91-98. 103, 106-107, 109-110, 112-113. June 1, 1970.

84. Bagge, Carl E. "Broadening the Supply Base of the Gas Industry." PUBLIC UTILITIES FORTNIGHTLY 85: 23-32. March 26, 1970.

85. Bagge, Carl E. "The Federal Power Commission." BOSTON COLLEGE INDUSTRIAL AND COMMERCIAL LAW REVIEW 11: 689-721. May 1970.

86. Bagge, Carl E. "Electric Power: An Industry in Crisis and Transition." PUBLIC UTILITIES FORTNIGHTLY 85: 55-65. June 4, 1970.

87. Bagge, Carl E. "The Quality for Life: Challenge to Regulation." PUBLIC UTILITIES FORTNIGHTLY 86: 15-20. September 10, 1970.

88. Bagge, Carl E. "One Dimensional Land Use Legislation: A Threat to Rational Energy Development." NATURAL RESOURCES LAWYER 6(74): Winter 1973.

89. Bailey, Richard. "Traditional Energy Resources: Present State and Future Development." FUTURES 4(2): 103-114. June 1972.

90. Bailey, Richard. "Britain and a Community Energy Policy." NATIONAL WESTMINSTER BANK QUARTERLY REVIEW: 5-15. November 1973.

91. Bailey, Richard. "The UK Coal Industry--Recent Past and Future." ENERGY POLICY 2(2): 152-158. June 1974.

92. Bailie, Richard C. and Alpert, Seymour. "Conversion of Municipal Waste to a Substitute Fuel." PUBLIC WORKS: August 1973.

93. Bakker-Arkema, F.W., et al. ENERGY IN MICHIGAN AGRICULTURE. Agricultural Engineering Department, Michigan State University, East Lansing, Michigan. 1974.

94. Baldwin, Malcolm F. "Public Policy on Oil--An Ecological Perspective." ECOLOGY LAW QUARTERLY 1: 245-303. Spring 1971.

95. Baldwin, Malcolm F. THE SOUTHWEST ENERGY COMPLEX: A POLICY EVALUATION. Conservation Foundation, Washington, D.C. 1973.

96. Balestra, Pietro. THE DEMAND FOR NATURAL GAS IN THE UNITED STATES. North-Holland, Amsterdam. 1967.

97. Ball, Norma R. "The East Coast of Scotland and North Sea Oil." GEOGRAPHY 58(258): 51-54. January 1973.

98. Balzhiser, Richard. "The Role of Nuclear Energy in Preserving Environmental Values." PUBLIC UTILITIES FORTNIGHTLY 91: 18-22. February 1, 1973.

99. Bandurski, Bruce Lord. "Ecology and Economics--Partners for Productivity." THE ANNALS OF THE AMERICAN ACADEMY OF POLITICAL AND SOCIAL SCIENCES 405: 75-94. January 1973.

100. Banner, Paul J. "The Energy Situation--A Rail Viewpoint." TRANSPORTATION JOURNAL 13(3): 15-19. Spring 1974.

101. BARCLAYS REVIEW. "The Balance of Power." BARCLAYS REVIEW 48(3): 53-56. August 1973.

102. BARCLAYS REVIEW. "Energy and the Common Market." BARCLAYS REVIEW 48(3): 60. August 1973.

103. BARCLAYS REVIEW. "The U.S. Oil Crisis." BARCLAYS REVIEW 48(3): 61. August 1973.

104. Barfield, Claude E. "Nuclear Power Plant Opponents Open Campaign to Halt Reactors." NATIONAL JOURNAL REPORTS 5(23): 850-851. June 9, 1973.

105. Barfield, Claude E. "Energy Research Funds." NATIONAL JOURNAL REPORTS 5(42): 1574. October 20, 1973.

106. Barfield, Claude E. "New Energy Offices Plan." NATIONAL JOURNAL REPORTS 6(12): 439. March 23, 1974.

107. Barfield, Claude E. "Energy Policy Organization." NATIONAL JOURNAL REPORTS 6(26): 962-967. June 29, 1974.

108. Barfield, Claude E. and Corrigan, Richard. "Environment Policy Law." NATIONAL JOURNAL 4(9): 336-349. February 26, 1972.

109. Barfield, Claude E., Corrigan, Richard and Noone, James A. "Developing Energy Policy." NATIONAL JOURNAL REPORTS 5(46): 1722-1730. November 17, 1973.

110. Barker, Mary L. "Oil Tankers and Canada's Pacific Seaboard." GEOSCIENCE CANADA 1: 31-34. 1974.

111. Barnes, Robert. "International Oil Companies Confront Governments: A Half-Century of Experience." INTERNATIONAL STUDIES QUARTERLY 16(4): 454-471. December 1972.

112. Barnett, Harold J. and Morse, Chandler. SCARCITY AND GROWTH: THE ECONOMICS OF NATURAL RESOURCE AVAILABILITY. Johns Hopkins Press, Baltimore. 1963.

113. Bass, D.M. (ed.). "State and Federal Regulations Pertaining to the Petroleum Industry." QUARTERLY OF THE COLORADO SCHOOL OF MINES 65: July 1970. (Whole Issue).

114. Bates, James Leonard. THE ORIGINS OF TEAPOT DOME: PROGRESSIVES, PARTIES, AND PETROLEUM, 1909-1912. University of Illinois Press, Urbana. 1963.

115. Battelle Memorial Institute. A REVIEW AND COMPARISON OF SELECTED U.S. ENERGY FORECASTS. Prepared for the Executive Office of the President, Office of Science and Technology, Energy Policy Staff. Battelle Memorial Institute, Washington, D.C. 1969.

116. Battelle Memorial Institute. A BRIEF OVERVIEW OF THE ENERGY REQUIREMENTS OF THE DEPARTMENT OF DEFENSE. Battelle Memorial Institute, Columbus, Ohio. 1972.

117. Battelle Energy Program. Energy Information Center. SUGGESTED ENERGY READINGS FROM THE ENERGY INFORMATION CENTER OF THE BATTELLE ENERGY PROGRAM. Battelle Energy Program, Battelle Memorial Institute, Washington, D.C. 1973.

118. BATTELLE RESEARCH OUTLOOK. "Our Energy Supply and Its Future." BATTELLE RESEARCH OUTLOOK 4(1): 1-31. 1972.

119. Bauer, John and Costello, Peter. PUBLIC ORGANIZATION OF ELECTRIC POWER: CONDITIONS, POLICIES AND PROGRAM. Harper, New York. 1949.

120. Baughman, Martin. "Energy Systems: Modeling and Policy Analysis." n.p. n.d.

121. Beane, Marjorie and Ross, John E. THE ROLE OF TECHNICAL INFORMATION IN DECISIONS ON NUCLEAR POWER PLANTS. IES Report 19. Institute for Environmental Studies, University of Wisconsin, Madison. (Forthcoming, September 1974).

122. Beizer, James and Kaufman, K.A. "Mining's Public Land Prospects Soar." IRON AGE 205: 49-54. June 25, 1970.

123. Bennett, Elmer. "Fuel Resources for the Future." THE CONFERENCE BOARD RECORD 9(7): 23. July 1972.

124. Berg, Charles A. "Energy Conservation Through Effective Utilization." SCIENCE 81: 128-138. July 13, 1973.

125. Berg, Charles A. "A Technical Basis for Energy Conservation." TECHNOLOGY REVIEW 76: 14-23. February 1974.

126. Berg, Charles A. "Conservation in Industry." SCIENCE 184(4134): 264-270. April 19, 1974.

127. Berg, Charles A. "Conservation is Effective Use of Energy at the Point of Consumption." In ENERGY: DEMAND, CONSERVATION AND INSTITUTIONAL PROBLEMS, Michael Macrakis (ed.). M.I.T. Press, Cambridge, Massachusetts. 1974.

128. Berg, R.R., Calhoun, J.C., Jr. and Whiting, R.L. "Progress for Expanded U.S. Production of Crude Oil." SCIENCE 184(4134): 331-335. April 19, 1974.

129. Bergen, G.S.P. "Some Thoughts on Power Supply and the Environment." RECORD OF THE ASSOCIATION OF THE BAR OF THE CITY OF NEW YORK 26: 677. November 1971.

130. Bergman, Roland. "Indian Subsistence in the Upper Amazon Rain Forest." Paper Presented at the Annual Meeting of the American Association for the Advancement of Science, San Francisco. February 1974.

131. Berkowitz, David A. and Squires, Arthur M. (eds.). POWER GENERATION AND ENVIRONMENTAL CHANGE. M.I.T. Press, Cambridge, Massachusetts. 1971.

132. Berman, M.B. and Hammer, M.J. THE IMPACT OF ELECTRICITY PRICE INCREASES ON INCOME GROUPS: A CASE STUDY IN LOS ANGELES. National Technical Information Service, Springfield, Virginia. March 1973.

133. Berman, M.B., Hammer, M.J. and Tihansky, D.P. THE IMPACT OF ELECTRICITY PRICE INCREASES ON INCOME GROUPS: WESTERN UNITED STATES AND CALIFORNIA. National Technical Information Service, Springfield, Virginia. November 1972.

134. Berner, Arthur S. and Scoggins, Sue. "Oil and Gas Drilling Programs -- Structure and Regulation." GEORGE WASHINGTON LAW REVIEW 41(3): 471-504. March 1973.

135. Berreby, Jean-Jacques. "Does America Need Arab Oil?" NEW MIDDLE EAST (19): 9-12. April 1970.

136. Berreby, Jean-Jacques. "Oil in the Orient: Syria's Offensive Against Saudi Arabia--U.S. Oil Link." NEW MIDDLE EAST (22): 12-13. July 1970.

137. Berreby, Jean-Jacques. "About That Arab Oil...Does the U.S. Really Need It?" ATLAS 19: 42-44. September 1970.

138. Berreby, Jean-Jacques. "A New Oil Strategy." THE ATLANTIC PAPERS (1): 47-57. 1972.

139. Berry, R. Stephen and Fels, Margaret Fulton. "The Production and Consumption of Automobiles: An Energy Analysis of the Manufacture, Discard and Reuse of the Automobile and its Component Materials." Department of Chemistry, University of Chicago. July 1972.

140. Berry, R. Stephen and Fels, Margaret Fulton. "The Energy Cost of Automobiles." SCIENCE AND PUBLIC AFFAIRS BULLETIN OF THE ATOMIC SCIENTISTS 29(10):11-17. December 1973.

141. Best, J.A. "Recent State Initiatives on Power Plant Siting: A Report and Comment." NATURAL RESOURCES LAWYER: Fall 1972.

142. Bickel, David and Markel, Clark. "Western North Dakota High School Senior Profiles on Coal Development." Cooperative Education Program and Experimental College, Minot State College, Minot, North Dakota. April 1974.

143. Bigda, Richard J. "Here's How Oil Planners Can Live With New Environmental Rules." OIL AND GAS JOURNAL 69: 66-68. July 19, 1971.

144. Binder, Denis. "A Novel Approach to Reasonable Regulation of Strip Mining." UNIVERSITY OF PITTSBURGH LAW REVIEW 34(3): 339-374. Spring 1973.

145. Biswas, Asit K. ENERGY AND THE ENVIRONMENT. Report Number 1. Planning and Finance Service, Environment Canada. Information Canada, Ottawa. n.d.

146. Bjornson, Bjorn. "The Challenge of Environmental Changes." PUBLIC UTILITIES FORTNIGHTLY 86: 13-18. August 13, 1970.

147. Blackett, George H. "Measuring Raw Materials Needs to the Year 2000." CONFERENCE BOARD RECORD 8: 23-28. January 1971.

148. Blaney, Harry C. "The Energy Crisis and International Cooperation." FOREIGN SERVICE JOURNAL 50: 12-14, 29-30. August 1973.

149. Blaney, Harry C. "The Energy Crisis: A Challenge to the International System." WORLD AFFAIRS 136(3): 195-207. Winter 1973-1974.

150. Boffey, Philip M. "Energy Crisis: Environmental Issue Exacerbates Power Supply Problem." SCIENCE: June 26, 1970.

151. Boffey, Philip M. "Radiation Standards: Are the Right People Making Decisions?" SCIENCE 171: 780-783. February 26, 1971.

152. Boland, John J., Geyer, John C. and Hanke, Steve H. ECONOMIC CONSIDERATIONS IN POWER PLANT SITING IN THE CHESAPEAKE BAY REGION: PHASE TWO REPORT. Department of Geography and Environmental Engineering, Johns Hopkins University, Baltimore, Maryland. March 1974.

153. Bonbright, James C. PUBLIC UTILITIES AND THE NATIONAL POWER POLITICS, Columbia University Press, New York. 1940.

154. Booz, Allen and Hamilton, Inc. AN INVENTORY OF ENERGY RESEARCH. National Science Foundation, Washington, D.C. 1971.

155. Boulding, Kenneth E. "A Look at National Priorities." CURRENT HISTORY 59: 65-72. August 1970.

156. Boulding, Kenneth E. "The Economics of Energy." ANNALS OF THE AMERICAN ACADEMY OF POLITICAL AND SOCIAL SCIENCE 410: 120-126. November 1973.

157. Boulding, Kenneth E. "The Social System and the Energy Crisis." SCIENCE 184(4134): 255-256. April 19, 1974.

158. Bowers, H.I. and Myers, M.L. ESTIMATED CAPITAL COSTS OF NUCLEAR AND FOSSIL POWER PLANTS. Oak Ridge National Laboratory, Oak Ridge, Tennessee. 1971.

159. Bowes, John and Stamm, Keith R. "Communication During Rapid Development of Energy Resources: A Coorientation Analysis." Paper Presented to the 1974 meeting of the International Communication Association. Communication Research Center, University of North Dakota, Grand Forks. April 1974.

160. Box, Thadis et al. REHABILITATION POTENTIAL OF WESTERN COAL LANDS. Ballinger Publishing Company, Cambridge, Massachusetts. 1974.

161. Boyd, F.C. "Nuclear Power in Canada: A Different Approach." ENERGY POLICY 2(2): 126-135. June 1974.

162. Bradley, Paul G. THE ECONOMICS OF CRUDE PETROLEUM PRODUCTION. North-Holland, Amsterdam. 1967.

163. Bradley, Paul G. "Increasing Scarcity: The Case of Energy Resources." AMERICAN ECONOMIC REVIEW 63(2): 119-125. May 1973.

164. Bradshaw, Thornton F. "Keeping the Energy Peace." VISTA 9(1): 20-23. August 1973.

165. Brady, David and Althoff, Phillip. "The Politics of Regulation: The Case of the Atomic Energy Commission and the Nuclear Industry." AMERICAN POLITICS QUARTERLY 1(3): 361-384. July 1973.

166. Branch, Melville C. "Oil Extraction, Urban Environment and City Planning." AIP JOURNAL: May 1972.

167. Brannon. "The Role of Taxes and Subsidies in United States Energy Policy." Draft report to the Energy Policy Project, Ford Foundation. 1971.

168. Braymer, Daniel T. THE CONTROLLED ENERGY GROWTH CONCEPT: THE NATIONAL ECONOMY AND PROSPECTS FOR POWER. National Electrical Manufacturers Association, New York. 1971.

169. Breyer, Stephen G. and MacAvoy, Paul W. ENERGY REGULATION BY THE FEDERAL POWER COMMISSION. n.p. n.d.

170. Breyer, Stephen and MacAvoy, Paul W. "The Natural Gas Shortage and the Regulation of Natural Gas Producers." HARVARD LAW REVIEW 86(6): 941-987. April 1973.

171. Bridges, Jack. "The National Energy Dilemma, Displayed for Total Viewing." THE CONFERENCE BOARD RECORD 10(8): 14-30. August 1973.

172. Bright, Arthur A. THE ELECTRIC LAMP INDUSTRY. Macmillan, New York. 194

173. Bright, James R. and Schoeman, Milton E.F. (eds.) A GUIDE TO PRACTICAL TECHNOLOGICAL FORECASTING. Prentice-Hall, Englewood Cliffs, New Jersey. 1973.

174. Brinkworth, B.J. SOLAR ENERGY FOR MAN. Halsted Press, New York. 1974.

175. British Petroleum Company. BP STATISTICAL REVIEW OF THE WORLD OIL INDUSTRY. British Petroleum Company, London. 1972.

176. Brobst, D.A. and Pratt, Walden P. (eds.). U.S. MINERAL RESOURCES. U.S. Geological Survey Professional Paper Number 820. U.S. Government Printing Office, Washington, D.C. 1973.

177. Brondel, G. "The Community's Problems in Petrol Supply." REVUE DU MARCHE COMMUN (165): 200-203. May 1973.

178. Bronstein, P.A. "AEC Decision-Making Process and the Environment: A Case Study of the Calvert Cliffs (Calvert Cliffs Coordinating Comm. Inc. V, U.S. Atomic Energy Comm. 449F2d1109) Nuclear Power Plant." ECOLOGY LAW QUARTERLY 1: 689. 1971.

179. Bronstein, P.A. "State Regulation of Power Plant Siting." ENVIRONMENTAL LAW 3: 273. 1973.

180. Brooks, David B. "Ocean Mining: The Political and Economic Aspects." MANAGEMENT REVIEW. November 1969.

181. Brooks, David B. and Krutilla, John V. PEACEFUL USE OF NUCLEAR EXPLOSIVES: SOME ECONOMIC ASPECTS. The Johns Hopkins Press, Baltimore. 1969.

182. Brown, A.E. and Berkowitz, E.B. "Energy Conservation at an Industrial Research Center." SCIENCE 184 (4134): 271-272. April 19, 1974.

183. Brown, Herbert H. "Utility Load Growth, the Environment, and FPC Responsibility." PUBLIC UTILITIES FORTNIGHTLY 85: 37-40. May 7, 1970.

184. Brown, Keith C. BIDDING FOR OFFSHORE OIL: TOWARD AN OPTIMAL STRATEGY. Southern Methodist University Press, Dallas. 1969.

185. Brown, Keith C. (ed.). REGULATION OF THE NATURAL GAS PRODUCING INDUSTRY. Resources for the Future, Washington, D.C. 1972.

186. Brubaker, Earl R. "Some Effect of Policy on Productivity in Soviet and American Crude Petroleum Extraction." JOURNAL OF INDUSTRIAL ECONOMICS 18: 33-52. November 1969.

187. Bruce, C.J. "The Open Petroleum Economy: A Comparison of Keynesian and Alternative Formulation." SOCIAL AND ECONOMIC STUDIES 21(2): 125-152. June 1972.

188. Brune, W.D., Jr. "The Economic Impact of Electric Power Development." Paper presented at the National Engineering Week Symposium, Chico State College, February 1972.

189. Bryan, R.M., Nichols, B.L. and Ramsey, J.N. SUMMARY OF RECENT LEGISLATIVE AND REGULATORY ACTIVITIES AFFECTING THE ENVIRONMENTAL QUALITY OF NUCLEAR FACILITIES. Oak Ridge National Laboratory, Oak Ridge, Tennessee. 1971.

190. Buchanan, J.R. "Nuclear Safety Information Center, its Products and Services." SPECIAL LIBRARIES 61: 492-495. November 1970.

191. Buchmann, A.P. "Electric Transmission and the Environment." CLEVELAND STATE LAW REVIEW 21: 121. 1972.

192. BULLETIN OF THE ATOMIC SCIENTISTS. "Energy Crisis; Symposium; Discussion: Part I." BULLETIN OF THE ATOMIC SCIENTISTS: March 1972.

193. BULLETIN OF THE ATOMIC SCIENTISTS. "Energy Crisis; Symposium; Discussion: Part II." BULLETIN OF THE ATOMIC SCIENTISTS 28(5): May 1972.

194. BULLETIN OF THE ATOMIC SCIENTISTS. "Energy and Environment." BULLETIN OF THE ATOMIC SCIENTISTS 28(5): 5-7. May 1972.

195. BULLETIN OF THE EUROPEAN COMMUNITIES. "Energy Policy: New Proposals to the Council." BULLETIN OF THE EUROPEAN COMMUNITIES 6(7-8): 17-21. 1973.

196. BULLETIN OF THE EUROPEAN COMMUNITIES. "The Community and the Energy Crisis." BULLETIN OF THE EUROPEAN COMMUNITIES 7(1): 10-16. 1974.

197. Bunyard, Peter. "The Power Crisis." ECOLOGIST 1: 4-7. October 1970.

198. Burch, W. and Bormann, F.H. (eds.). GROWTH LIMITS AND THE QUALITY OF LIFE. W.H. Freeman, San Francisco. 1974.

199. Burchard, Hans-Joachim. "Towards a Common European Energy Policy." AUSSENPOLITIK 21(1): 76-82. 1970.

200. Burg, Nan C. ENERGY CRISIS IN THE UNITED STATES: A SELECTED BIBLIOGRAPHY OF NON-TECHNICAL MATERIALS. Exchange Bibliography Number 550. Council of Planning Librarians, Monticello, Illinois. 1974.

201. Burn, Duncan L. THE POLITICAL ECONOMY OF NUCLEAR ENERGY: AN ECONOMIC STUDY OF CONTRASTING ORGANIZATIONS IN THE UK AND USA, WITH EVALUATION OF THEIR EFFECTIVENESS. Research Monograph Number 9. Institute of Economic Affairs, London. 1967.

202. Burrows, James C. and Domencich, Thomas A. AN ANALYSIS OF THE UNITED STATES OIL IMPORT QUOTA. Heath Lexington Books, Lexington, Massachusetts. 1970.

203. BUSINESS WEEK. "Why the Oil Giants Are Under the Gun: Special Report." BUSINESS WEEK: 82-83. October 25, 1969.

204. BUSINESS WEEK. "The Politics Behind the New Oil Hunt." BUSINESS WEEK (2166): 104-106. March 6, 1971.

205. Byerton, Gene. NUCLEAR DILEMMA. Ballantine, New York, 1970.

206. Caldwell, Lynton Keith. "Coal and Institutional Arrangements." In SYMPOSIUM ON COAL AND PUBLIC POLICIES, HELD AT THE UNIVERSITY OF TENNESSEE, KNOXVILLE, OCTOBER 13-15 1971, F. Schmidt-Bleek and R.S. Carlsmith (eds.). Center for Business and Economic Research, College of Business Administration, The University of Tennessee, Knoxville, Tennessee. 1972. (pp. 149-163).

207. Caldwell, Lynton Keith. "A National Policy for Energy?" INDIANA UNIVERSITY LAW REVIEW 47(2): 624-635. Summer 1972.

208. Caldwell, Lynton Keith. "Energy and Environment: The Bases for Public Policies." ANNALS OF THE AMERICAN ACADEMY OF POLITICAL AND SOCIAL SCIENCE 410: 127-138. November 1973.

209. Caldwell, Lynton Keith. "Energy and the Structure of Social Institutions." Paper Presented at the Annual Meeting of the American Association for the Advancement of Science, San Francisco. February 1974.

210. California Institute of Technology. Environmental Quality Laboratory. PEOPLE, POWER AND POLLUTION: ENVIRONMENTAL AND PUBLIC INTEREST ASPECTS OF ELECTRIC POWER PLANT SITING. California Institute of Technology, Pasadena. 1971.

211. CALIFORNIA JOURNAL. "California and the Energy Crisis." CALIFORNIA JOURNAL: June 1973.

212. Callahan, John C. and Callahan, Jacqueline G. EFFECTS OF STRIP MINING AND TECHNOLOGICAL CHANGE ON COMMUNITIES AND NATURAL RESOURCES IN INDIANA'S COAL MINING.REGION. Research Bulletin Number 871. Agricultural Experiment Station, Purdue University, Lafayette, Indiana. January 1971.

213. Cambel, Ali B. "Impact of Energy Demands." PHYSICS TODAY 23: 38-43, 45. December 1970.

214. Cambel, Ali B. et al. ENERGY R & D AND NATIONAL PROGRESS. Prepared for the Interdepartmental Energy Study by the Energy Study Group. U.S. Government Printing Office, Washington, D.C. 1965.

215. Campbell, M. Earl. "The Energy Outlook for Transportation in the United States." TRAFFIC QUARTERLY 22(2): 183-210. April 1973.

216. Campbell, Penelope N. (ed.). THE SANTA BARBARA OIL SPILL: SOME PROBLEMS IN HISTORICAL INTERPRETATION. Mimeo Report for Environmental Studies 4. University of California, Santa Barbara. Spring 1973.

217. Campbell, Robert W. THE ECONOMICS OF SOVIET OIL AND GAS. Published for Resources for the Future by the Johns Hopkins Press, Baltimore. 1968.

218. Canada National Energy Board. ENERGY SUPPLY AND DEMAND IN CANADA AND EXPORT DEMAND FOR CANADIAN ENERGY 1966 to 1990. Information Canada, Ottawa. 1969.

219. CANADIAN BUSINESS. "The Best of Two Worlds: How the National Oil Policy Could Change." CANADIAN BUSINESS 46(4): 14-19. April 1973.

220. Capron, W.M. (ed.). TECHNOLOGIC CHANGE IN REGULATED INDUSTRIES. Brookings Institution, Washington, D.C. 1971.

221. Carbon, M.W. and Houlberg, W.A. NUCLEAR POWER AND THE ENVIRONMENT: PROCEEDINGS OF A STUDENT CONFERENCE HELD AT MADISON, WISCONSIN, APRIL 3 AND 4, 1970. University of Wisconsin Press, Madison. 1970.

222. Carey, Jane Perry Clark. "Iran and Control of Its Oil Resources." POLITICAL SCIENCE QUARTERLY 89(1): 147-174. March 1974.

223. Carlsmith, R.S. et al. "Electrical Energy and its Environmental Impact." In ORNL-NSF ENVIRONMENTAL PROGRAM PROGRESS REPORT -- JUNE 30, 1972. Oak Ridge National Laboratory, Oak Ridge, Tennessee. January 1973. (pp. 5-21).

224. Carpenter, Richard A. "Technology Assessment and the Congress." In TECHNOLOGY ASSESSMENT: UNDERSTANDING THE SOCIAL CONSEQUENCES OF TECHNOLOGICAL APPLICATION, Raphael G. Kasper (ed.). Praeger, New York. 1969.

225. Carter, A.P. "Applications of Input-Output Analysis to Energy Problems." SCIENCE 184(4134): 325-330. April 19, 1974.

226. Carter, A.P. (ed.). STRUCTURAL INTERDEPENDENCE, ENERGY AND THE ENVIRONMENT. University Press of New England, Hanover, New Hampshire. (Forthcoming).

227. Carter, Luther J. "Warm-Water Irrigation: An Answer to Thermal Pollution?" SCIENCE 165: 478-480. August 1, 1969.

228. Carter, Luther J. "Land Use: Congress Taking Up Conflict Over Power Plants." SCIENCE 170: 718-719. November 13, 1970.

229. Carter, Luther J. "Rio Blanco: Stimulating Gas and Conflict in Colorado." SCIENCE 180: 844. 1973.

230. Carter, Luther J. "Deepwater Ports: Issue Mixes Supertankers, Land Policy." SCIENCE 181:825. 1973.

231. Carter, Luther J. "Florida: An Energy Policy Emerges in a Growth State." SCIENCE 184(4134): 302-306. April 19, 1974.

232. Carver, John A., Jr. "The Lawyer's Concerns With the Energy Crisis." NATURAL RESOURCES LAWYER 6(4): 553-562. Fall 1973.

233. Case, Clifford and Schoenbrod, David. "Electricity or the Environment: A Study of Public Regulation Without Public Control." CALIFORNIA LAW REVIEW 61: 961-1010. June 1973.

234. CASE WESTERN RESERVE LAW REVIEW. "U.S. -- Canadian Energy Resource Development." CASE WESTERN RESERVE LAW REVIEW 5(36). Winter 1972.

235. Casey, W.J. "Department Discusses International Ramifications of the Energy Situation." Statement May 1, 1973. U.S. DEPARTMENT OF STATE BULLETIN 68: 702-706. May 28, 1973.

236. Casey, W.J. "International Cooperation on Energy." Address June 21, 1973. U.S. DEPARTMENT OF STATE BULLETIN 69: 59-62. July 9, 1973.

237. Cassuto, A. "Competition in Nuclear Energy: International Issues." WORLD TODAY 20: 277-284. July 1964.

238. Caudill, Harry Monroe. NIGHT COMES TO THE CUMBERLANDS: A BIOGRAPHY OF A DEPRESSED AREA. Little, Brown, Boston. 1963.

239. Caudill, Harry Monroe. MY LAND IS DYING. E.P. Dutton and Company, Inc., New York. 1971.

240. Cavers, D.F. "Administrative Decision-Making in Nuclear Facility Licensing." UNIVERSITY OF PENNSYLVANIA LAW REVIEW 110: 330-370. January 1962.

241. Cederstam, Lennart. "Nuclear Power in Sweden, Planning for the 1980's." EUROPEAN FREE TRADE ASSOCIATION BULLETIN 15(2): 15-16. March 1974.

242. Center for Economic and Social Information. "Oil and the Poor Countries." ENVIRONMENT 16(2): 10-14. March 1974.

243. Cetron, Marvin J. and Coates, Vary T. "Energy and Society." PROCEEDINGS OF THE ACADEMY OF POLITICAL SCIENCE 31(2): 33-40. December 1973.

244. Chamber of Commerce of the United States of America. Community and Regional Development Group. MEETING NATIONAL ENERGY NEEDS. Chamber of Commerce of the United States of America, Washington, D.C. 1972.

245. Chandler, Geoffrey. "The Myth of Oil Power: International Groups and National Sovereignty." INTERNATIONAL AFFAIRS (London) 46: 710-718. October 1970.

246. Chapman, Duane, Tyrrell, Timothy and Mount, Timothy. "Electricity Demand Growth and the Energy Crisis." SCIENCE 178(4062): 703-708. November 17, 1972.

247. Chapman, Duane, Tyrell, Timothy and Mount, Timothy. "More Resistance to Electricity." ENVIRONMENT 15(8): 18-36. October 1973.

248. Chapman, Peter F. "Energy Production -- A World Limit?" NEW SCIENTIST 47: 634-636. September 24, 1970.

249. Chapman, Peter F. "Energy Costs -- A Review of Methods." ENERGY POLICY 2(2): 91-103. June 1974.

250. Charles River Associates. AN ANALYTICAL FRAMEWORK FOR EVALUATING THE OIL IMPORT QUOTA PROGRAM. Prepared for Energy Policy Staff, Office of Science and Technology, Executive Office of the President. Charles River Associates, Cambridge, Massachusetts. 1969.

251. Chase Manhattan Bank. FUTURE GROWTH OF THE WORLD PETROLEUM INDUSTRY. Chase Manhattan Bank, New York. 1961.

252. Chase Manhattan Bank. CAPITAL INVESTMENTS OF THE WORLD PETROLEUM INDUSTRY. Chase Manhattan Bank, New York. 1967.

253. Chase Manhattan Bank. OUTLOOK FOR ENERGY IN THE UNITED STATES TO 1985. Chase Manhattan Bank, New York. 1972.

254. Chase Manhattan Bank. Energy Division. CAPITAL INVESTMENT OF THE WORLD PETROLEUM INDUSTRY, 1970. Chase Manhattan Bank, New York. 1971.

255. Cheaney, Edgar S. and Eibling, James A. "Reducing the Consumption of Energy." BATTELLE RESEARCH OUTLOOK 4(1): 1972.

256. CHEMICAL & ENGINEERING NEWS. "Soviet Union Faces 1970 Energy Deficit." CHEMICAL & ENGINEERING NEWS 47: 28-30. November 24, 1969.

257. Chersky, Nikolai, Makogon, Yuri and Belov, V. "Solid Gas -- World Reserves Are Enormous." OIL AND GAS INTERNATIONAL 10: 82-84, 89-90. August 1970.

258. Christian, Virgil L., Jr. and Vaughan, Claude M. "Some Aspects of Cost and Demand in the Pricing of Electric Power." LAND ECONOMICS 47(3): 281-288. August 1971.

259. CHRISTIAN CENTURY. "Politics of Energy." CHRISTIAN CENTURY 40: 222. February 21, 1973.

260. Cicchetti, Charles J. ALASKAN OIL: ALTERNATIVE ROUTES AND MARKETS. Published for Resources for the Future by the Johns Hopkins University Press, Baltimore. 1973.

261. Cicchetti, Charles J. "The Wrong Route." ENVIRONMENT 15(3): 4-13. June 1973.

262. Cicchetti, Charles J. and Gillen, William J. "The Mandatory Oil Import Quota Program: A Consideration of Economic Efficiency and Equity." NATURAL RESOURCES JOURNAL 13(3): 399-430. July 1973.

263. Ciriacy-Wantrup, Siegfried von. "The Economics of Environmental Policy." LAND ECONOMICS 47: 36-45. February 1971.

264. Citizens' Advisory Committee on Environmental Quality. CITIZEN ACTION GUIDE TO ENERGY CONSERVATION. U.S. Government Printing Office, Washington, D.C. 1973.

265. Clark, R.B. "Organization Against Oil." NEW SCIENTIST 43: 658-660. September 25, 1969.

266. Clark, Timothy B. "Gas Allocation Politics." NATIONAL JOURNAL REPORTS 6(11): 397- . March 16, 1974.

267. Coe, Robert N. "Power and the Environment." PLANNING: February 1973.

268. Cohen, Arlan A., et al. "Asthma and Air Pollution from a Coal-Fueled Power Plant." AMERICAN JOURNAL OF PUBLIC HEALTH 62(9): 1181-1188. September 1972.

269. Colorado School of Mines. Potential Gas Agency. POTENTIAL SUPPLY OF NATURAL GAS IN THE UNITED STATES AS OF DECEMBER 31, 1966. Colorado School of Mines, Golden. 1967.

270. Columbia University, Graduate School of Business. ENERGY AND MAN: A SYMPOSIUM. Appleton-Century-Crofts, New York. 1960.

271. Comar, C.L. "Biological Aspects of Nuclear Weapons." AMERICAN SCIENTIST 50: 339-353. 1962.

272. Commoner, Barry. THE CLOSING CIRCLE. Alfred Knopf, New York. 1971.

273. Commoner, Barry, Corr, Michael and Stamler, Paul J. "The Cause of Pollution." ENVIRONMENT: April 1971.

274. CONGRESSIONAL DIGEST. "Controversy Over the Impact of Federal Regulations on the Petroleum Situation." CONGRESSIONAL DIGEST 52(10): 227-233. October 1973.

275. CONGRESSIONAL DIGEST. "PRO and CON Discussion: Would Relaxing the Present Degree of Federal Regulation of the U.S. Oil Industry Help Ease the 'Energy Crisis'?" CONGRESSIONAL DIGEST 52(10): 234-251. October 1973.

276. CONGRESSIONAL QUARTERLY. "Energy-Environment Dilemma: National Policy Sought." CONGRESSIONAL QUARTERLY WEEKLY REPORT 30: 1018-1023. May 6, 1972.

277. CONGRESSIONAL QUARTERLY. "El Paso Pipeline Fight: A Race in Congress and the Courts." CONGRESSIONAL QUARTERLY: September 2, 1972.

278. CONGRESSIONAL QUARTERLY. "Energy Crisis: Fact of Life for Millions in 1973." CONGRESSIONAL QUARTERLY: February 3, 1973.

279. CONGRESSIONAL QUARTERLY. "Energy Crisis in America. CONGRESSIONAL QUARTERLY: March 1973. Special Issue.

280. CONGRESSIONAL QUARTERLY. "Off-shore Drilling, Oil Quotas and Rationing." CONGRESSIONAL QUARTERLY: March 31, 1973.

281. CONGRESSIONAL QUARTERLY. "End Import Quotas, Decontrol Gas Prices." CONGRESSIONAL QUARTERLY: April 21, 1973.

282. CONGRESSIONAL QUARTERLY. "Power Problems: Consumption Up, Supplies Short." CONGRESSIONAL QUARTERLY: September 18, 1973.

283. CONGRESSIONAL QUARTERLY. "Energy: Conflict Ahead on Allocation Fuels." CONGRESSIONAL QUARTERLY: October 6, 1973.

284. Connery, Robert H. and Gilmour, Robert S. (eds.). THE NATIONAL ENERGY PROBLEM. Academy of Political Science, New York. 1974.

285. Connor, James E. "Prospects for Nuclear Power." PROCEEDINGS OF THE ACADEMY OF POLITICAL SCIENCE: 63-73. December 1973.

286. Connor, John T. "Onward into Overdevelopment." THE CONFERENCE BOARD RECORD 10(8): 10-13. August 1973.

287. Cook, C. Sharp. "Energy: Planning for the Future." AMERICAN SCIENTIST 61(1): 61-65. January-February 1973.

288. Cook, Donald C. "Capability of Electric Utilities to Fulfill Future Needs." PUBLIC UTILITIES FORTNIGHTLY 95: 17-23. July 16, 1970.

289. Cook, Donald C. "Is an Energy Crisis Inevitable?" PUBLIC UTILITIES FORTNIGHTLY: August 16, 1973.

290. Cook, Earl. "The Flow of Energy in an Industrial Society." SCIENTIFIC AMERICAN 224(3): 134-147. September 1971.

291. Cook, Earl. "Energy Sources for the Future." THE FUTURIST 6(4): 142-150. August 1972.

292. Cook, Franklin H. "Public Ownership: United Kingdom Versus United States." PUBLIC UTILITIES FORTNIGHTLY 84: 26-34. December 18, 1969.

293. Cook, Franklin H. "Is Public Power the Answer?" PUBLIC UTILITIES FORTNIGHTLY 85: 20-30. February 1970.

294. Cook, Michael C. ENERGY IN THE UNITED STATES 1960-1985: ANALYSIS AND FORECAST OF CONSUMPTION PATTERNS FOR FUELS AND ELECTRICITY. Sartorius, New York. 1967.

295. Cook, Thomas J. and Scioli, Frank P., Jr. "Obstacles to Evaluating the Impact of Social Action Policy." Paper Presented at the American Society for Public Administration Action, Hearing on Federal Delivery Systems. New York. March 1972.

296. Cook, Thomas J. and Scioli, Frank P., Jr. "A Research Strategy for Analysing the Impacts of Public Policy." ADMINISTRATIVE SCIENCE QUARTERLY: September 1972.

297. Cook, Thomas J. and Scioli, Frank P., Jr. "Experimental Design in Policy Impact Analysis." SOCIAL SCIENCE QUARTERLY: September 1973.

298. Cooper, D.F. "The Changing Pattern of Energy Purchasing in Europe." PURCHASING JOURNAL 27: 32-37. July 1970.

299. Cootner, Paul H. and Löf, George O.G. WATER DEMAND FOR STEAM ELECTRIC GENERATION: AN ECONOMIC PROJECTION MODEL. The Johns Hopkins Press, Baltimore. 1966.

300. Copeland, Otis L. and Packer, Paul E. "Land Use Aspects of the Energy Crisis and Western Mining." JOURNAL OF FORESTRY 70(11): 671-675. November 1972.

301. Corbett, Lindsay. "North Sea Oil -- Information Requirements and Supply." THE INFORMATION SCIENTIST 7(3): 111-112. September 1973.

302. Corden, Max. "Implications of the Oil Price Rise." JOURNAL OF WORLD TRADE LAW 8(2): 133-143. March-April 1974.

303. Corr, M. and MacLeod, D. "Getting It Together: Energy Use in Communes and Other Alternative Life Styles." ENVIRONMENT: November 1972.

304. Corrigan, Richard. "Offshore Leasing." NATIONAL JOURNAL 4(28): 1109-1116. July 8, 1972.

305. Corrigan, Richard. "Oil Imports." NATIONAL JOURNAL 4(35): 1357-1366. August 26, 1972.

-22-

306. Corrigan, Richard. "Energy Policy." NATIONAL JOURNAL 4(43): 1621-1631. October 21, 1972.

307. Corrigan, Richard. "El Paso Gas Ventures Affect U.S. Policies, Consumer Prices." NATIONAL JOURNAL 5(3): 67-75. January 20, 1973.

308. Corrigan, Richard. "Administration Reviews Ban on Santa Barbara Oil Drilling." NATIONAL JOURNAL 5(15): 547. April 14, 1973.

309. Corrigan, Richard. "Washington Report: US Energy Policy After the President's Message." ENERGY POLICY 1(1): 65-70. June 1973.

310. Corrigan, Richard. "Controlling Oil Allotments." NATIONAL JOURNAL REPORTS 5(34): 1262-1276. August 25, 1973.

311. Corrigan, Richard. "Alaskan Oil, Anti-Trust and the President's Second Message." ENERGY POLICY 1(2): 161-163. September 1973.

312. Corrigan, Richard. "Elk Hills Oil Reserve." NATIONAL JOURNAL REPORTS 5(49): 1839- . December 8, 1973.

313. Corrigan, Richard. "Washington Report: The Real Crisis and Self-Sufficiency by 1980." ENERGY POLICY 2(1): 71-72. March 1974.

314. Corrigan, Richard. "Washington Report: The National Academy of Sciences' Forum on Energy." ENERGY POLICY 2(2): 165-166. June 1974.

315. Corrigan, Richard and Barfield, Claude E. "Alaska Pipeline Lobbying." NATIONAL JOURNAL REPORTS 5(32): 1172-1178. August 11, 1973.

316. Cotterill, Ewan M.R. "The Northwest Territories Face New Challenges." THE CONFERENCE BOARD RECORD 10(8): 33-38. August 1973.

317. Cottrell, W.B. SOME CONSIDERATIONS REGARDING ATMOSPHERIC POLLUTION FROM POWER PRODUCTION. Oak Ridge National Laboratory, Oak Ridge, Tennessee. 1970.

318. Cottrell, William Frederick. ENERGY AND SOCIETY: THE RELATION BETWEEN ENERGY, SOCIAL CHANGE, AND ECONOMIC DEVELOPMENT. McGraw-Hill, New York. 1955. Reprinted by Greenwood Press, Westport, Connecticut, 1970.

319. Council on Economic Priorities. THE PRICE OF POWER: ELECTRIC UTILITIES AND THE ENVIRONMENT. The Council on Economic Priorities, Washington, D.C. 1972.

320. Council on Economic Priorities. "Leased and Lost: A Study of Public and Indian Coal Leasing in the West." ECONOMIC PRIORITIES REPORT 5(2): 1974.

321. Crabb, Cecil V., Jr. "The Energy Crisis, the Middle East, and American Foreign Policy." WORLD AFFAIRS 136(1): 48-73. Summer 1973.

322. Craig, James B. "Power to the People." AMERICAN FORESTS 78( ): 36-39, 56. April 1972.

323. Crane, David. "Canada's Energy Policies in a Global Context." INTERNATIONAL PERSPECTIVES: 32-37. July-August 1973.

324. Crossland, Janice. "Cars, Fuel, and Pollution." ENVIRONMENT 16(2): 15-27. March 1974.

325. Culbertson, O.L. THE CONSUMPTION OF ELECTRICITY IN THE UNITED STATES. Oak Ridge National Laboratory, Oak Ridge, Tennessee. 1971.

326. Cunningham, James P. AN ENERGETIC MODEL LINKING FOREST INDUSTRY AND ECOSYSTEMS. Communication 79.3. Finnish Forest Research Institute, Helsinki. 1974.

327. Curtis, Carl T. "The Energy Crisis -- Real or Imagined?" NEBRASKA MUNICIPAL REVIEW: April 1971.

328. Curtis, Richard and Hogan, Elizabeth. PERILS OF THE PEACEFUL ATOM: THE MYTH OF SAFE NUCLEAR POWER PLANTS. Doubleday, Garden City, New York. 1969.

329. Czapowskyj, Miroslaw M. "Anthracite Coal-Mine Spoils Today -- Forests Tomorrow." PENNSYLVANIA FORESTS 59: 81-83. Fall 1969.

330. Dalsted, Norman L. et al. "Economic Impact of Alternative Energy Development Patterns in North Dakota." Department of Agricultural Economics, North Dakota State University, Fargo. April 1974.

331. Dalsted, Norman L., Leistritz, F. Larry and Hertsgaard, Thor A. "Energy Resources Development in the Northern Great Plains: A Summary of Economic Impacts." Department of Agricultural Economics, North Dakota State University, Fargo. April 1974.

332. Darmstadter, Joel. ENERGY IN THE WORLD ECONOMY. Published for Resources for the Future by the Johns Hopkins Press, Baltimore. 1971.

333. Darmstadter, Joel. "Energy." In POPULATION, RESOURCES AND THE ENVIRONMENT: THE COMMISSION ON POPULATION GROWTH AND THE AMERICAN FUTURE RESEARCH REPORTS VOLUME 3, Ronald G. Ridker (ed.). U.S. Government Printing Office, Washington, D.C. 1972. (pp. 103-149).

334. Darmstadter, Joel. "Energy Consumption: Trends and Patterns." In ENERGY, ECONOMIC GROWTH AND THE ENVIRONMENT, Sam H. Schurr (ed.). Johns Hopkins Press, Baltimore. 1972.

335. Darmstadter, Joel. "Energy Consumption: Trends, Prospects and Issues." JOURNAL OF SOIL AND WATER CONSERVATION 28( ): 108-113. May-June 1973.

336. Darmstadter, Joel. "World Energy Requirements." In ENERGY, THE ENVIRONMENT, AND HUMAN HEALTH, Asher J. Finkel (ed.). Publishing Sciences Group, Acton, Massachusetts. 1973.

337. Darmstadter, Joel and Hunter, Robert E. "Energy in Crisis?" In THE UNITED STATES AND THE DEVELOPING WORLD: AGENDA FOR ACTION, 1973. Overseas Development Council, Washington, D.C. 1973. (pp. 90-100).

338. Darmstadter, Joel, Tietlebaum, Perry D. and Polach, Jaroslav. ENERGY IN THE WORLD'S ECONOMY: A STATISTICAL REVIEW OF TRENDS IN OUTPUT, TRADE AND CONSUMPTION SINCE 1925. Johns Hopkins Press, Baltimore. 1971.

339. Davidson, Paul. "The Depletion Allowance Revisited." NATURAL RESOURCES JOURNAL 10: 1-9. January 1970.

340. Davis, Frank W., Jr. and Malcolm, John G. "Will Evolving National Issues Create a Rail Renaissance?" MSU BUSINESS TOPICS 21(1): 37-46. Winter 1973.

341. Davis, K.C. "Nuclear Facilities Licensing, Another View." UNIVERSITY OF PENNSYLVANIA LAW REVIEW 110( ): 371- . January 1962.

342. Dean, Flora. A BIBLIOGRAPHY OF NON-TECHNICAL LITERATURE ON ENERGY. U.S. Government Printing Office, Washington, D.C. 1971.

343. Dean, Genevieve C. "Energy in the People's Republic of China." ENERGY POLICY 2(1): 33-54. March 1974.

344. Debeboise, T.M. and Madden, W.J., Jr. "Impact of the National Environmental Policy Act Upon Administration of the Federal Power Act." LAND AND WATER LAW REVIEW 8( ): 93- . 1973.

345. De Bell, Garrett. "Energy." In ENVIRONMENTAL HANDBOOK, Ballantine, New York. 1970.

346. Dee, Norbert, et al. "Financing Abatement of Mine Drainage Pollution: Case Study, Appalachia." WATER RESOURCES BULLETIN 8(3): 473-482. June 1972.

347. Degler, Stanley E. (ed.). OIL POLLUTION: PROBLEMS AND POLICIES. Environmental Mangement Series. BNA Books, Washington, D.C. 1969.

348. Degolyer and MacNaughton. REPORT ON NATIONAL EMERGENCY POLICY. U.S. Office of Naval Petroleum and Oil Shale Reserves. U.S. Government Printing Office, Washington, D.C. 1971.

349. DeNike, L.D. "Radioactive Malevolence." SCIENCE AND PUBLIC AFFAIRS: 16. February 1974.

350. Denton, Jesse C. AN ASSESSMENT OF THE NATIONAL ENERGY PROBLEM. National Science Foundation, Washington, D.C. 1971.

351. Denton, Jesse C. (ed.). GEOTHERMAL ENERGY. University of Alaska Press, Fairbanks. 1972.

352. Denver University. Future Requirements Committee. Denver Research Institute. FUTURE NATURAL GAS REQUIREMENTS OF THE UNITED STATES. Denver University, Denver. 1969.

353. Desprairies, Pierre. L'Evolution de la Crise Petroliere de 1970-1971." REVUE DE DEFENSE NATIONALE 28( ): 138-773. May 1972.

354. Dhrymes, Phoebus J. and Kurz, Mordecai. "Technology and Scale in Electricity Generation." ECONOMETRICA 32 ( ): 287-315. July 1964.

355. DiBona, Charles J. "Administration Policies Affecting the Natural Gas Industry." NATURAL RESOURCES LAWYER 6(4): 503-512. Fall 1973.

356. DiBona, Charles J. "Reconciling Our Energy and Environmental Demands." In ENERGY AND THE ENVIRONMENT: A COLLISION OF CRISES, Irwin Goodwin (ed.). Publishing Sciences Group, Acton, Massachusetts. 1973.

357. Dienes, Leslie. "Geographical Problems of Allocation in the Soviet Fuel Supply." ENERGY POLICY 1(1): 3-20. June 1973.

358. Dinsmore, John H. "The Energy Crunch." LIBRARY JOURNAL 99(9): 1270-1273. May 1, 1974.

359. Ditton, Robert B. and Goodale, Thomas L. (eds.). ENVIRONMENTAL IMPACT ANALYSIS: PHILOSOPHY AND METHODS. Sea Grant Publications Office, University of Wisconsin, Madison. 1972.

360. Doctor, Ronald D. "Growth and Energy Demands." Paper Presented at the Annual Meeting of the American Association for the Advancement of Science, San Francisco. February 1974.

361. Doctor, Ronald D. et al. CALIFORNIA'S ELECTRICITY QUANDARY: III. SLOWING THE GROWTH RATE. National Technical Information Service, Springfield, Virginia. September 1972.

362. Doub, William O. FEDERAL ENERGY REGULATION: AN ORGANIZATIONAL STUDY. Federal Energy Regulation Study Team, Washington, D.C. 1974.

363. Downs, James F. "The Social Consequences of a Dry Well." AMERICAN ANTHROPOLOGIST 67( ): 1387-1416. 1965.

364. DRAKE LEGAL REVIEW. "Power Plant Siting -- a Regulatory Crisis." DRAKE LEGAL REVIEW: June 1973.

365. Drechsler, Herbert D. "Exploitation of the Continental Margin." PROCEEDINGS OF THE ACADEMY OF POLITICAL SCIENCE 31(2): 98-110. December 1973.

366. Drefus, Daniel A. FEDERAL ENERGY ORGANIZATION. A Staff Analysis of the Senate Committee on Interior and Insular Affairs. U.S. Government Printing Office, Washington, D.C. 1973.

367. Dubin, Fred S. "Energy Conservation Needs New Architecture and Engineering." PUBLIC POWER: March/April 1972.

368. Ducaesneau. "Competition in the Energy Industry." Draft report to the Energy Policy Project, Ford Foundation. n.d.

369. Due, John F. "The Developing Economies, Tax and Royalty Payments by the Petroleum Industry, and the United States Income Tax." NATURAL RESOURCES JOURNAL 10: 10-26. January 1970.

370. Dugger, Ronnie. "Oil and Politics." ATLANTIC 224( ): 66-90. September 1969.

371. Dunlavy, D.C. "Government Regulation of Atomic Energy." UNIVERSITY OF PENNSYLVANIA LAW REVIEW 105( ): 295- . January 1957.

372. Dupree, Walter G., Jr. and West, James A. UNITED STATES ENERGY THROUGH THE YEAR 2000. U.S. Government Printing Office, Washington, D.C. 1972.

373. Dwivedi, O.P. (ed.). PROTECTING THE ENVIRONMENT: ISSUES AND CHOICES -- CANADIAN PERSPECTIVES. Copp Clark, Toronto. 1974.

374. Eaton, E.D. "Power Plant Siting: The Energy-Environment Dilemma." WATER SPECTRUM 4( ): 36-40. Summer 1972.

375. Ebasco Services, Inc. ENERGY CONSUMPTION AND SUPPLY TRENDS CHART BOOK. Ebasco Services, Inc. 1970.

376. Ebbin, Steven and Kasper, Raphael. CITIZEN GROUPS AND THE NUCLEAR POWER CONTROVERSY: USES OF SCIENTIFIC AND TECHNOLOGICAL INFORMATION. M.I.T. Press, Cambridge, Massachusetts. 1974.

377. Eckles, Robert B. "What Now -- Federal Power Commission?" PUBLIC UTILITIES FORTNIGHTLY 85: 33-37. June 18, 1970.

378. Eckstein, Otto. WATER RESOURCE DEVELOPMENT: THE ECONOMICS OF PROJECT EVALUATION. Harvard University Press, Cambridge, Massachusetts. 1958.

379. ECOLOGY LAW QUARTERLY. "Taxation as a Tool of Natural Resource Management: Oil as a Case Study." ECOLOGY LAW QUARTERLY: Fall 1971.

380. ECONOMIC BULLETIN, NATIONAL BANK OF EGYPT. "Arab Oil and Energy Supplies." ECONOMIC BULLETIN, NATIONAL BANK OF EGYPT 26(3): 221-225. 1973.

381. THE ECONOMIST. Intelligence Unit. SOVIET OIL TO 1980. ECONOMIST, London. n.d.

382. ECONOMIST (eds.). "The Oil Wealth." THE ECONOMIST 247(6767): 39-45. 1973.

383. Edel, Mathew. "Autos, Energy, and Pollution." ENVIRONMENT 15(8): 10-17. October 1973.

384. Edel, Mathew. "Autos Feed on Oil." ENVIRONMENT 15(9): 34-37. November 1973.

385. Edison Electric Institute. QUESTIONS AND ANSWERS ABOUT THE ELECTRIC UTILITY INDUSTRY. Edison Electric Institute, New York. n.d.

386. Edison Electric Institute. ENERGY ECONOMICS AND THE ENVIRONMENT. Edison Electric Institute, New York. 1969.

387. Edison Electric Institute. BIBLIOGRAPHY AND DIGEST OF U.S. ELECTRIC AND TOTAL ENERGY FORECASTS, 1970-2050. Edison Electric Institute, New York. 1970.

388. Edison Electric Institute. Committee on Environment. Plant Siting Task Force. MAJOR ELECTRIC POWER FACILITIES AND THE ENVIRONMENT. Edison Electric Institute, Chicago. 1970.

389. Educational Facilities Laboratory. TOTAL ENERGY: A TECHNICAL REPORT. Educational Facilities Laboratory, New York. 1967.

390. Educational Foundation for Nuclear Science. "The Energy Crisis: Part I." BULLETIN OF THE ATOMIC SCIENTISTS 27(7): 2-53. September 1971.

391. Educational Foundation for Nuclear Science. "The Energy Crisis: Part II." BULLETIN OF THE ATOMIC SCIENTISTS 27(8): 2-56. October 1971.

392. Educational Foundation for Nuclear Science. "The Energy Crisis: Part III." BULLETIN OF THE ATOMIC SCIENTISTS 27(9): 38-56. November 1971.

393. Edwards, Richard W., Jr. "Many Splendored Possibilities or Hobson's Choice? -- Who Made the Policies and What are the Assumptions." CASE WESTERN RESERVE JOURNAL OF INTERNATIONAL LAW 5(1): 39-51. Winter 1972.

394. Efford, Ian E. and Smith, Barbara M. (eds.). ENERGY AND THE ENVIRONMENT: H.R. MACMILLAN LECTURES FOR 1971. Institute of Resource Ecology, University of British Columbia, Vancouver. 1972.

395. Egerton, Alfred. "Civilization and the Use of Energy." ADVANCEMENT OF SCIENCE 7: 386-397. March 1951.

396. Ehrlich, Paul R. and Holden, J.P. "Energy Crisis." SATURDAY REVIEW 54: 50-51. August 7, 1971.

397. Eipper, A.W. "Pollution Problems, Resource Policy, and the Scientist." SCIENCE 189( ): 11-15. July 3, 1970.

398. Eipper, A.W., Carlson, C.A. and Hamilton, L.S. "Impacts of Nuclear Power Plants on the Environment." LIVING WILDERNESS 34: 5-12. Autumn 1970.

399. Eisenbud, Merril. "Health Hazards From Radioactive Emissions." In ENERGY, THE ENVIRONMENT, AND HUMAN HEALTH, Asher J. Finkel (ed.). Publishing Sciences Group, Acton, Massachusetts. 1973.

400. Eisenbud, Merril and Gleason, George. ELECTRIC POWER AND THERMAL DISCHARGES. Gordon and Breach, New York. 1969.

401. EKISTICS. "Energy Resources, Human Comfort and the Environment." EKISTICS: May 1972.

402. Eklund, Sigvard. "Atomic Energy -- Key to Industrial Progress." INTERECONOMICS (7): 206-209. July 1969.

403. Ekmann, James. "An Energy Policy." In ARIZONA ENERGY: A RESOURCE CATALOG, Helen Ingram, Hanna Cortner and John Dettloff (eds.). Institute of Government Research, University of Arizona, Tucson. n.d. (pp. III-1 -- III-33).

404. Elder, J.A., Durer, E.J. and Gulish, W.J. "On the Accident Proneness of the World Tanker Fleet and Its Implication Vis-a-Vis Pollution." Gulf Oil Corporation. April 1971.

405. Electric Research Council. ELECTRIC UTILITIES INDUSTRY RESEARCH AND DEVELOPMENT GOALS THROUGH THE YEAR 2000. Electric Research Council, Washington, D.C. 1971.

406. Electric Utility Industry Task Force on Environment. THE ELECTRIC UTILITY INDUSTRY AND THE ENVIRONMENT: A REPORT TO THE CITIZENS ADVISORY COMMITTEE ON RECREATION AND NATURAL BEAUTY. New York. 196

407. Ellingen, Dana C. and Towsey, William E. A BIBLIOGRAPHY OF CONGRESSIONAL PUBLICATIONS ON ENERGY FROM THE 89TH CONGRESS TO JULY 1, 1971. Prepared at the request of Henry M. Jackson, Chairman, Committee on Interior and Insular Affairs, U.S. Senate, pursuant to S. Res. 45, a National Fuels and Energy Policy Study. U.S. Government Printing Office, Washington, D.C. 1971.

408. Ellingen, Dana C. A SUPPLEMENTAL BIBLIOGRAPHY OF PUBLICATIONS ON ENERGY. U.S. Government Printing Office, Washington, D.C. 1972.

409. Ellingen, Dana C. RESOLVED: THAT THE FEDERAL GOVERNMENT SHOULD CONTROL THE SUPPLY AND UTILIZATION OF ENERGY IN THE UNITED STATES: A PRELIMINARY BIBLIOGRAPHY OF MATERIALS RELATING TO THE INTERCOLLEGIATE DEBATE TOPIC, 1973-1974. U.S. Library of Congress, Washington, D.C. July 1973.

410. Elliott, John M. "Environmental Aspects of Nuclear Power." URBAN LAWYER 4( ): 33-38. Winter 1972.

411. Elliott, Martin A. THE ROLE OF GAS IN MEETING FUTURE ENERGY DEMANDS. Institute of Gas Technology, Chicago. 1959.

412. Ellison, S.P. TOWARDS A NATIONAL POLICY ON ENERGY RESOURCES AND MINERAL PLANT FOODS. Special Report FLDIGP 21D2A. University of Texas, Austin. December 1973.

413. El Mallakh, Ragaei. SOME DIMENSIONS OF MIDDLE EAST OIL: THE PRODUCING COUNTRIES AND THE UNITED STATES. American-Arab Association for Commerce and Industry, New York. 1970.

414. Ely, Northcutt. SUMMARY OF MINING AND PETROLEUM LAWS OF THE WORLD. U.S. Government Printing Office, Washington, D.C. 1970.

415. Ely, Northcutt. "Resolution Re Energy Problems and Report." NATURAL RESOURCES LAWYER 6(4): 565-572. Fall 1973.

416. Endahl, Lowell. "Rural Electric Industrial Development Programs." MANAGEMENT QUARTERLY 13(3): 21-24. Fall 1972.

417. ENERGY: DEMAND, CONSERVATION, AND INSTITUTIONAL PROBLEMS. Proceedings of a conference held at the Massachusetts Institute of Technology, February 12-14, 1973. M.I.T. Press, Cambridge, Massachusetts. 1973.

418. Energy and Life Symposium, Kellogg Center, Michigan State University, March 1, 1974. Proceedings in Press.

419. ENERGY POLICY. "Energy Modelling and Policy Making." ENERGY POLICY 2 (1): 2. March 1974.

420. ENERGY POLICY. "Energy Supplies -- A Temporary Crisis or a Permanent Problem? A Discussion." ENERGY POLICY 2(1): 67-70. March 1974.

421. ENERGY POLICY (eds.). "Energy Budgets and Rational Planning." ENERGY POLICY 2(2): 90. June 1974.

422. Energy Policy Project. STATE AND LOCAL DECISION-MAKING ON ENERGY POLICY. The Energy Policy Project, Washington, D.C. 1971.

423. Energy Research Unit, Queen Mary College, London. "World Energy Modelling: The Development of Western European Oil Prices." ENERGY POLICY 1(1): 21-34. June 1973.

424. Engler, Robert. THE POLITICS OF OIL: A STUDY OF PRIVATE POWER AND DEMOCRATIC DIRECTIONS. Macmillan, New York. 1961.

425. Enk, Gordon A. BEYOND NEPA: CRITERIA FOR ENVIRONMENTAL IMPACT REVIEW. Institute on Man and Science, Rensselaerville, New York. May 1973.

426. Enos, John L. PETROLEUM PROGRESS AND PROFITS: A HISTORY OF PROCESS INNOVATION. M.I.T. Press, Cambridge, Massachusetts. 1962.

427. ENVIRONMENT: "Low Energy Living." ENVIRONMENT 15(3): 33-35. April 1973

428. Environment Information Center, Inc. THE ENERGY INDEX: A SELECT GUIDE TO ENERGY INFORMATION SINCE 1970. Environment Information Center, Inc., New York. 1973.

429. ENVIRONMENTAL SCIENCE AND TECHNOLOGY. "Geothermal Heats Up." ENVIRONMENTAL SCIENCE AND TECHNOLOGY: August 1973.

430. Erickson, Edward W. ECONOMIC INCENTIVES, INDUSTRIAL STRUCTURE AND THE SUPPLY OF CRUDE OIL DISCOVERIES IN THE UNITED STATES, 1946-1958-59. Unpublished Ph.D. dissertation, Vanderbilt University, Nashville, Tennessee. 1968.

431. Erickson, Edward W. "Crude Oil Prices, Drilling Incentives and the Supply of New Discoveries." NATURAL RESOURCES JOURNAL 10: 27-52. January 1970.

432. Evan, Harry Z. "The Multinational Oil Company and the Nation State." JOURNAL OF WORLD TRADE LAW 4: 666-685. September-October 1970.

433. Evensen, Jens. "The North Sea Oil." INTERNASJONAL POLITIKK (2): 359-373. April-June 1973.

434. Evensen, Jens. "The North Sea Oil Problems of International Law and International Politics." INTERNASJONAL POLITIKK (3): 501-514. July-September 1973.

435. Fabricant, Neil and Hallman, Robert M. TOWARDS A RATIONAL NATIONAL POWER POLICY: ENERGY, POLITICS AND POLLUTION. A Report of the Environmental Protection Agency of the City of New York. George Braziller, Inc., New York. 1971.

436. Facca, G. and Ten Dam, A. GEOTHERMAL POWER ECONOMICS. Worldwide Geothermal Exploration Co., Los Angeles. 1964.

437. Falcon, Norman. "Oil in its True Proportions." THE GEOGRAPHICAL MAGAZINE 45(3): 187-191. December 1972.

438. Falls, O.B. "A Survey of the Market for Nuclear Power in Developing Countries." ENERGY POLICY 1(3): 225-242. December 1973.

439. Faltermayer, Edmund. "The Energy 'Joyride' is Over." FORTUNE 86(3): 99-102. September 1972.

440. Fanning, Leonard M. THE SHIFT OF WORLD PETROLEUM POWER AWAY FROM THE UNITED STATES. Gulf Oil Corp., Pittsburgh. 1958.

441. Farber, Erich A. "Solar Energy." In ENERGY, THE ENVIRONMENT, AND HUMAN HEALTH, Asher J. Finkel (ed.). Publishing Sciences Group, Acton, Massachusetts. 1973.

442. Farber, John P. and Newton, Charles G. "Anticipated Energy Resources Development Impact on High School Youth in Converse County, Wyoming." Office of State Federal Relations, State of Wyoming. April 1974.

443. Federal Bar Association. Committee on Federal Utility and Power Law. FEDERAL STATUTES RELATING TO ELECTRIC POWER AND NATIONAL ELECTRIC POWER POLICY. Federal Bar Association, New York. 1967.

444. Feehan, John G. "The Energy Crisis and the Consumer States." NATURAL RESOURCES LAWYER 6(4): 495-502. Fall 1973.

445. Fehd, Carolyn S. "Productivity in the Petroleum Pipelines Industry." MONTHLY LABOR REVIEW: April 1971.

446. Felson, Marcus. "The Prospective Role of Sociology in Understanding the Energy Problem." Department of Sociology, University of Illinois, Urbana. 1974.

447. Field, Michael. "Oil: OPEC and Participation." THE WORLD TODAY 28(1): 5-13. January 1972.

448. Finkel, Asher J. ENERGY, THE ENVIRONMENT, AND HUMAN HEALTH. Publishing Sciences Group, Acton, Massachusetts. 1973.

449. Finon, Dominique. "Optimisation Model for the French Energy Sector." ENERGY POLICY 2(2): 136-151. June 1974.

450. Fischer, Loyd and Biere, Arlo (eds.). ENERGY AND AGRICULTURE: RESEARCH IMPLICATIONS. Report Number 2. North Central Research Strategy Committee on Natural Resource Development. October 1973.

451. Fischman, Leonard L. and Landsberg, Nans H. "Adequacy of Nonfuel Minerals and Forest Resources." In POPULATION, RESOURCES AND THE ENVIRONMENT: THE COMMISSION ON POPULATION GROWTH AND THE AMERICAN FUTURE RESEARCH REPORTS VOLUME 3, Ronald G. Ridker (ed.). U.S. Government Printing Office, Washington, D.C. 1972. (pp. 77-101).

452. Fisher, Anthony C. et al. "The Economics of Environmental Preservation: A Theoretical and Empirical Analysis." AMERICAN ECONOMIC REVIEW: September 1972.

453. Fisher, Franklin M. SUPPLY AND COSTS IN THE U.S. PETROLEUM INDUSTRY: TWO ECONOMETRIC STUDIES. Published for Resources for the Future by the Johns Hopkins Press, Baltimore. 1964.

454. Fisher, Franklin M. and Kaysen, Carl. A STUDY IN ECONOMETRICS: THE DEMAND FOR ELECTRICITY IN THE UNITED STATES. North-Holland, Amsterdam. 1962.

455. Fisher, John C. ENERGY CRISES IN PERSPECTIVE. Wiley-Interscience, New York. 1974.

456. Fitch, James B. THOUGHTS FOR THE ENERGY CRISIS: THE ECONOMICS OF INSULATION AND HEATING SYSTEMS IN TYPICAL WILLIAMETTE VALLEY HOMES. Man and His Activities as Related to Environmental Quality. Study Number 2. Department of Agricultural Economics, Oregon State University, Corvallis. January 1974.

457. Flawn, Peter T. ENVIRONMENTAL GEOLOGY. CONSERVATION, LAND-USE PLANNING AND RESOURCE MANAGEMENT. Harper, New York. 1970.

458. Foell, W.K. et al. 1973 SURVEY OF ENERGY USE IN WISCONSIN. IES Report Number 10. Institute for Environmental Studies, University of Wisconsin, Madison. 1974.

459. Fogel, Robert W. RAILROADS AND AMERICAN ECONOMIC GROWTH. Johns Hopkins Press, Baltimore. 1964.

460. Folk, Hugh and Hannon, Bruce. AN ENERGY, POLLUTION, AND EMPLOYMENT POLICY MODEL. CAC Document Number 68. Center for Advanced Computations, University of Illinois, Urbana. n.d.

461. FORBES. "Energy Yardsticks of Management Performance." FORBES 107: 144, 146-147. January 1, 1971.

-33-

462. FORBES. "Energy: A Lot of Growth Problems Add Up to an Uncertain Future." FORBES 111(1): 106-140. January 1, 1973.

463. FORBES. "The Energy Crisis." FORBES 112(11): 27. December 1, 1973.

464. FORBES. "Energy." FORBES 113(1): 207-212. January 1, 1974.

465. Ford, Daniel F. et al. THE NUCLEAR FUEL CYCLE, A SURVEY OF THE PUBLIC HEALTH, ENVIRONMENTAL AND NATIONAL SECURITY EFFECTS OF NUCLEAR POWER. Union of Concerned Scientists, Cambridge, Massachusetts. 1973.

466. Ford, Daniel F., and Kendall, Henry W. "Nuclear Safety." ENVIRONMENT 14(7): 2-9. September 1972.

467. Ford Foundation. Energy Policy Project. EXPLORING ENERGY CHOICES. Ford Foundation, Washington, D.C. 1974.

468. Foreman, Harry. NUCLEAR POWER AND THE PUBLIC. University of Minnesota Press, Minneapolis. 1970.

469. FOREST NOTES. "Nuclear Power Plans for New Hampshire: A Symposium." FOREST NOTES (100): 2-10. Fall 1969.

470. FORTUNE. "Energy." FORTUNE 85: 68-69. April 1972.

471. FORTUNE. "Energy: A Policy to Avert a Crisis." FORTUNE 86(3): 81-84. September 1972.

472. FORTUNE. "Evading the 'Arab Oil Squeeze'." FORTUNE 88(4): 126. October 1973.

473. FORTUNE. "Learning to Live with the Oil Squeeze." FORTUNE 88(6): 25-42. December 1973.

474. FORTUNE. "The Market System and the Energy Crisis." FORTUNE 88(6): 79-80. December 1973.

475. FORTUNE. "The Energy Crunch and National Leadership." FORTUNE 89(1): 65-68. January 1974.

476. Foster Associates. THE ROLE OF PETROLEUM AND NATURAL GAS FROM THE OUTER CONTINENTAL SHELF IN THE NATIONAL SUPPLY OF PETROLEUM AND NATURAL GAS. U.S. Government Printing Office, Washington, D.C. 1968.

477. Foster Associates, Inc. AN ANALYSIS OF THE REGULATORY ASPECTS OF NATURAL GAS SUPPLY. National Technical Information Service, Springfield, Virginia. 1973.

478. Fowler, Henry G. "Capital Supplies for Energy Supply." THE CONFERENCE BOARD RECORD 11(5): 23-25. May 1974.

479. Fowlkes, Frank V. and Havemann, Joel. "The Policies of Energy." NATIONAL JOURNAL REPORTS 5(49): 1830-1838. December 8, 1973.

480. Fox, Irving K. "Institutional Mechanisms." In SUMMARY REPORT OF THE CORNELL WORKSHOP ON ENERGY AND THE ENVIRONMENT, National Science Foundation. U.S. Government Printing Office, Washington, D.C. 1972.

481. Fraize, W.E. and Dukowicz, J.K. TRANSPORTATION ENERGY AND ENVIRONMENTAL ISSUES. Mitre Corporation, Washington, D.C. 1972.

482. Frank, Helmut J. CRUDE OIL PRICES IN THE MIDDLE EAST: A STUDY IN OLIGOPOLISTIC PRICE BEHAVIOR. Praeger, New York. 1966.

483. Frank, Helmut J. "Economic Strategy for Import-Export Controls on Energy Materials." SCIENCE 184(4134): 316-320. April 19, 1974.

484. Frank, Helmut J. and Schanz, John H., Jr. "Natural Gas in the Future Energy Pattern." In REGULATION OF THE NATURAL GAS PRODUCING INDUSTRY, K.C. Brown (ed.). Johns Hopkins Press, Baltimore. 1970.

485. Frank, Helmut J. and Schanz, John J., Jr. "The Future of American Oil and Natural Gas." ANNALS OF THE AMERICAN ACADEMY OF POLITICAL AND SOCIAL SCIENCE 410( ): 24-34. November 1973.

486. Frank, Helmut J. and Wells, Donald A. "United States Oil Imports: Implications for the Balance of Payments." NATURAL RESOURCES JOURNAL 13(3): 431-447. July 1973.

487. Frankel, Herbert R. "Private vs. Public Power." CHALLENGE (NEW YORK) 4( ): 48-52. August-September 1956.

488. Frankel, Paul H. OIL: THE FACTS OF LIFE. Weidenfeld and Nicolson, London. 1962.

489. Frankel, Paul H. ESSENTIALS OF PETROLEUM: A KEY TO OIL ECONOMICS. Cass, London. 1969.

490. Frawley, Margaret. SURFACE MINED AREAS: CONTROL AND RECLAMATION OF ENVIRONMENTAL DAMAGE: A BIBLIOGRAPHY. National Technical Information Service, Springfield, Virginia. 1971.

491. Frazier, Charles H. and Stelzer, Irwin M. "Competitive Rate Making in Gas Distribution." PUBLIC UTILITIES FORTNIGHTLY 84: 42-47. October 23, 1969.

492. Freeman, S. David. "A New Form of Competition." PUBLIC POWER: May 1970.

493. Freeman, S. David. "Toward a Policy of Energy Conservation." BULLETIN OF THE ATOMIC SCIENTISTS 27(8): 8-12. October 1971.

494. Freeman, S. David. "The Energy Crisis: Is it Real?" In ENERGY AND THE ENVIRONMENT: A COLLECTION OF CRISES, Irwin Goodwin (ed.). Publishing Sciences Group, Washington, D.C. 1973.

495. Freeman, S. David. "Blame Uncle Sam -- Not Mother Nature -- for Fuel Shortages." PUBLIC POWER: March/April 1973.

496. Freeman, S. David. "The Energy Joyride is Over." SCIENCE AND PUBLIC AFFAIRS BULLETIN OF THE ATOMIC SCIENTISTS 29(8): 39. October 1973.

497. Freeman, S. David. "Is There an Energy Crisis? An Overview." ANNALS OF THE AMERICAN ACADEMY OF POLITICAL AND SOCIAL SCIENCE 410( ): 1-10. November 1973.

498. Freeman, S. David. "National Energy Policy." Twentieth Century Fund. 1974.

499. French, Cecil L. "Attitudes of Johnson County, Wyoming, Residents Toward Selected Aspects of Their Environment." Lakehead University, Thunder Bay, Ontario, Canada. April 1974.

500. Frey, J.W. and Ide, H.C. A HISTORY OF THE PETROLEUM ADMINISTRATION FOR WAR, 1941-1945. Petroleum Administration for War, Washington, D.C. 1946.

501. Friedlander, Gordon D. "New Directions in Power." MODERN GOVERNMENT: September 1972.

502. Friedlander, Gordon D. "Energy: Crisis and Challenge." IEEE SPECTRUM 10(5): 18-27. May 1973.

503. Friedlander, Gordon D. "Fuel/Energy Crisis: Toward a National Energy Policy." IEEE SPECTRUM 10(6): 36-43. June 1973.

504. Fulda, Michael. OIL AND INTERNATIONAL RELATIONS: ENERGY TRADE, TECHNOLOGY, AND POLITICS. Unpublished Ph.D. dissertation. American University, Washington, D.C. 1970.

505. Fuller, B. "Energy: Past and Future." EKISTICS 34(203): 241-245. October 1972.

506. Furlong, David B. "Bilateral Exploitation of North American Energy Resources -- An Introduction." CASE WESTERN RESERVE JOURNAL OF INTERNATIONAL LAW 5(1): 36-38. Winter 1972.

507. Future Requirements Agency. FUTURE NATURAL GAS REQUIREMENTS OF THE UNITED STATES. Denver Research Institute, Denver. 1971.

508. FUTURES. "Energy, Man and the Environment." FUTURES 4(2): 188-190. June 1972.

509. THE FUTURIST. "Coal: An Ancient Fuel with a Bright Future." THE FUTURIST 6(4): 151. August 1972.

510. THE FUTURIST. "The Depletion of Fossil Fuels." THE FUTURIST 6(4): 152. August 1972.

511. Gadda, David G. "Taxation as a Tool of Natural Resource Management: Oil as a Case Study." ECOLOGY LAW QUARTERLY 1( ): 749-772. Fall 1971.

512. Gage, Stephen J. "Who Should Pay for Clean Energy Research?" In ENERGY AND THE ENVIRONMENT: A COLLISION OF CRISES, Irwin Goodwin (ed.). Publishing Sciences Group, Acton, Massachusetts. 1973.

513. Gaither, Richard W. "Energy: Complex Target for Today's Technology." Address Delivered at the 3rd Urban Technology Conference, Boston, Massachusetts. September 26, 1973. National Bureau of Standards, Gaithersburg, Maryland. 1973.

514. Galatin, Malcolm. ECONOMICS OF SCALE AND TECHNICAL CHANGE IN THERMAL POWER GENERATION. North-Holland, Amsterdam. 1968.

515. Galway, Michael A. "A Continental Energy Policy -- An Examination of Some of the Current Issues." CASE WESTERN RESERVE JOURNAL OF INTERNATIONAL LAW 5(1): 65-80. Winter 1972.

516. Gardner, Stephen L. "Natural Gas -- Its Impending Shortage and Potential Abundance." FEDERAL RESERVE BANK OF DALLAS BUSINESS REVIEW: 1-5. January 1971.

517. Garfield, Paul J. and Lovejoy, Wallace F. PUBLIC UTILITY ECONOMICS. Prentice-Hall, Englewood Cliffs, New Jersey. 1964.

518. Garvey, Gerald. ENERGY, ECOLOGY, ECONOMY: A FRAMEWORK FOR ENVIRONMENTAL POLICY. W.W. Norton and Company, Inc., New York. 1972.

519. Garvin, D.F. "Social, Economic, and Utility Growth." PUBLIC UTILITIES FORTNIGHTLY 91( ): 23-38. February 15, 1973.

520. Gas Dynamics Symposium, 7th, Evanston, Illinois, 1968. ENERGY; PROCEEDINGS OF THE SEVENTH BIENNIAL GAS DYNAMICS SYMPOSIUM. Northwestern University Press, Evanston, Illinois. 1968.

-37-

521. Gaskin, Maxwell. "The Economic Impact of North Sea Oil on Scotland." THE THREE BANKS REVIEW (97): 30-50. March 1973.

522. Gatchell, Willard W. "The Portent of Changes in Attitudes in FPC Regulation." PUBLIC UTILITIES FORTNIGHTLY 95: 13-22. June 18, 1970.

523. Gavett, Earle E. "Agriculture: Energy Use and Conservation." Paper Presented in a seminar series on "Energy and Agriculture," Texas A & M University, College Station, Texas. 1972.

524. Gavett, Earle E. "Agriculture and the Energy Crisis." Paper Presented at the National Conference on Agriculture and Energy Crisis, University of Nebraska. April 1973.

525. Gehman, Clayton. U.S. ENERGY SUPPLIES AND USES. Staff Economic Study, Federal Reserve System Board of Governors. U.S. Government Printing Office, Washington, D.C. 1973.

526. GEOGRAPHICAL MAGAZINE. "World Map of the Energy Market." THE GEOGRAPHICAL MAGAZINE 46(8): 408. May 1974.

527. Georgescu-Roegen, Nicholas. THE ENTROPY LAW AND THE ECONOMIC PROBLEM. Distinguished Lecture Series Number 1. Department of Economics and Graduate School of Business, University of Alabama. December 1970.

528. Georgescu-Roegen, Nicholas. THE ENTROPY LAW AND ECONOMIC PROCESS. Harvard University Press, Cambridge, Massachusetts. 1971.

529. Georgescu-Roegen, Nicholas. "Energy and Economic Myths." In GROWTH LIMITS AND THE QUALITY OF LIFE, W. Burch and F.H. Bormann (eds.). W.H. Freeman, San Francisco. 1974.

530. Georgetown University. Center for Strategic and International Studies. UNDERSTANDING THE "NATIONAL ENERGY DILEMMA": A REPORT OF THE JOINT COMMITTEE ON ATOMIC ENERGY. Center for Strategic and International Studies, Georgetown University, Washington, D.C. 1973.

531. George Washington University. LEGAL, ECONOMIC AND TECHNICAL ASPECTS OF LIABILITY AND FINANCIAL RESPONSIBILITY AS RELATED TO OIL POLLUTION. Publication Number PB-198776. U.S. Department of Commerce, Washington, D.C. 1970.

532. Gerber, Abraham. "Environment and the Energy Industries." BUSINESS ECONOMICS: January 1971.

533. Gerber, Abraham. "Energy Growth and the Environment." PUBLIC UTILITIES FORTNIGHTLY 89(12): 69-74. June 8, 1972.

534. Gill, Crispin, Booker, F. and Soper, T. THE WRECK OF THE TORREY CANYON. Taplinger, New York. 1967.

535. Gillette, H.G. "Energy Policy -- Phase II." SCIENCE: July 13, 1973.

536. Gillette, Robert. "Nuclear Safety (I): The Roots of Dissent." SCIENCE 177( ): 771- . 1972.

537. Gillette, Robert. "Nuclear Safety (II): Years of Delay." SCIENCE 177( ): 867- . 1972.

538. Gillette, Robert. "Nuclear Safety (III): Critics Charge Conflicts of Interest." SCIENCE 177( ): 970- . 1972.

539. Gillette, Robert. "Nuclear Safety (IV): The Barriers to Communications." SCIENCE 177( ): 1080-1082. September 22, 1972.

540. Gillette, Robert. "Nuclear Safety: Atomic Energy Commission Report Makes the Best of It." SCIENCE 179( ): 360-363. January 26, 1973.

541. Gillette, Robert. "Fiscal 1974 Budget: Energy." SCIENCE 179: 549-550. February 9, 1973.

542. Gillette, Robert. "White House Energy Policy: Who Has the Power?" SCIENCE 179: 1211-1212. March 23, 1973.

543. Gillette, Robert. "Energy: The Muddle at the Top." SCIENCE 182( ): 1319-1321. December 28, 1973.

544. Gillette, Robert. "Budget: Energy." SCIENCE 183( ): 636-638. February 15, 1974.

545. Gillette, Robert. "Energy Reorganization: Progress in the Offing." SCIENCE 184(4135): April 26, 1974.

546. Gilmour, Robert S. "Political Barriers to a National Policy." PROCEEDINGS OF THE ACADEMY OF POLITICAL SCIENCE 31(2): 183- . December 1973.

547. Glaeser, Martin G. "A Critique of Public Power Policies." AMERICAN ECONOMIC REVIEW 47( ): 394-402. May 1957.

548. Glaser, Peter E. THE ENVIRONMENTAL CRISIS IN POWER GENERATION AND POSSIBLE FUTURE DIRECTIONS. Arthur D. Little, Cambridge, Massachusetts. 1971.

549. Gofman, John William and Tamplin, A.R. POISONED POWER: THE CASE AGAINST NUCLEAR POWER PLANTS. Rodale, Emmaus, Pennsylvania. 1971.

550. Gold, Raymond. "Social Impacts of Strip Mining and Other Industrializations of Coal Resources." Institute of Social Science Research, University of Montana, Missoula. n.d.

551. Goldberg, Michael A. "The Economics of Limiting Energy Use." ALTERNATIVES 1( ): 3-13. Summer 1972.

552. Goldman, Marvin. "Comparison Scales for Radioactive and Nonradioactive Hazards." In ENERGY, THE ENVIRONMENT, AND HUMAN HEALTH, Asher J. Finkel (ed.). Publishing Sciences Group, Acton, Massachusetts. 1973.

553. Goldsmith, John R. "Health Hazards From Chemical and Particulate Effluents." In ENERGY, THE ENVIRONMENT, AND HUMAN HEALTH, Asher J. Finkel (ed.). Publishing Sciences Group, Acton, Massachusetts. 1973.

554. Golze, Alfred R. "Impact of Urban Planning on Electric Utilities." PROFESSIONAL ENGINEER 43( ): 33-35. March 1973.

555. Gonchar, V.I. "Electric Power in the Southern Part of the Central Economic Region." SOVIET GEOGRAPHY: REVIEW AND TRANSLATION 15(3): 135-141. March 1974.

556. Gonzales, Ronald R. "Curtailment: Increased FPC Regulation of Direct Sales of Natural Gas." LOUISIANA LAW REVIEW 33(2): 335-338. Winter 1973.

557. Goodwin, Irwin. (ed.). ENERGY AND THE ENVIRONMENT: A COLLISION OF CRISES. Publishing Sciences Group, Acton, Massachusetts. 1973.

558. Gordon and Breach. ELECTRIC POWER AND THERMAL DISCHARGES. Atomic Industrial Forum, New York. 1969.

559. Gordon, R.R. "Developments in European and World Markets for Coking Coal and Cokes, ECE, Rome." ENERGY POLICY 1(1): 76. June 1973.

560. Gordon, Richard L. THE EVOLUTION OF ENERGY POLICY IN WESTERN EUROPE: THE RELUCTANT RETREAT FROM COAL. Praeger, New York. 1971.

561. Gordon, Richard L. "Alternatives to Oil and Natural Gas." PROCEEDINGS OF THE ACADEMY OF POLITICAL SCIENCE 31(2): 74-86. December 1973.

562. Gordon, Suzanne. BLACK MESA: ANGEL OF DEATH. John Day Company, New York. 1973.

563. Goss, W.P. and McGowan, J.G. "Transportation and Energy -- A Future Confrontation." TRANSPORTATION 1(3): 265-290. November 1972.

564. Goth, Harvey L. "Significant Developments in Federal Power Commission Pipeline Certificate Cases in 1968." NATURAL RESOURCES LAWYER 2: 208-220. July 1969.

565. GOVERNMENT EXECUTIVE. "The Federal Government and the Energy Crisis." GOVERNMENT EXECUTIVE 5(2): 39-40. February 1973.

566. Gracer, David A. and Schafer, Jack D. "Oil and Gas in the Seventies: Will There be Enough?" COLUMBIA JOURNAL OF WORLD BUSINESS 6( ): 59-68. May-June 1971.

567. Grafftey, Heward. "We Need an Energy Policy -- Now!" CANADIAN BUSINESS 47(1): 22-29. January 1974.

568. Grass, Vivien C. "State Regulation of Power Plant Siting." INDIANA LAW JOURNAL 47(4): Summer 1972.

569. Great Britain. Ministry of Power. FUEL POLICY. Her Majesty's Stationery Office, London. 1967.

570. Green, Harold P. "Nuclear Technology and the Fabric of Government." THE GEORGE WASHINGTON LAW REVIEW 33: 121-161. October 1964.

571. Green, Harold P. "Safety Determination in Nuclear Power Licensing: A Critical View." NOTRE DAME LAW 43( ): 633- . June 1968.

572. Green, Harold P. "Nuclear Safety and the Public Interest." Text of Address Delivered at Nuclear Safety Program Annual Information Meeting, Oak Ridge, Tennessee. February 1972.

573. Green, Harold P. "Nuclear Power: Risk, Liability, and Indemnity." MICHIGAN LAW REVIEW 71(3): 479-510. January 1973.

574. Green, Harold P. "The Great Delusion: Public Participation in Nuclear Power Licensing." Manuscript submitted to WILLIAM AND MARY LAW REVIEW for Publication in their Symposium on Nuclear Power. May 1974

575. Gregory, Derek P. "Hydrogen Energy Systems." In ENERGY, THE ENVIRONMENT, AND HUMAN HEALTH, Asher J. Finkel (ed.). Publishing Sciences Group, Acton, Massachusetts. 1973.

576. Grimes, John A. "Wanted: A Rational Energy Policy." THE AMERICAN FEDERATIONIST 80(4): 6-11. April 1973.

577. Grot, Richard and Socolow, Robert H. "Energy Utilization in a Residential Community." Paper Presented at M.I.T. Conference, February 1973.

578. Grot, Richard A. and Socolow, Robert H. "Energy Utilization in Residential Community." In ENERGY: DEMAND, CONSERVATION AND INSTITUTIONAL PROBLEMS, Michael Macrakis (ed.). M.I.T. Press, Cambridge, Massachusett 1974.

579. Gulf Oil Corporation. UNITED STATES PRIMARY ENERGY DEMAND, 1950-1985: BASIC DATA AND FORECASTS. Gulf Oil Corporation, Pittsburgh. 1968.

580. Gumpel, Werner. "Soviet Oil and Soviet Middle East Policy." AUSSEN POLITIK 23(1): 104- . 1972.

581. Gustafson, Philip F. "Nuclear Power and Thermal Pollution: Zion, Illinois." BULLETIN OF THE ATOMIC SCIENTISTS 26: 17-23. May 1970.

582. Gustavson, Marvin. DIMENSIONS OF WORLD ENERGY. Mitre Corporation, Washington, D.C. 1971.

583. Guyol, Nathaniel B. THE WORLD ELECTRIC POWER INDUSTRY. University of California Press, Berkeley. 1969.

584. Guyol, Nathaniel B. ENERGY IN THE PERSPECTIVE OF GEOGRAPHY. Prentice-Hall, Englewood Cliffs, New Jersey. 1971.

585. Hafele, W. "Energy Choices That Europe Faces: A European View of Energy." SCIENCE 184(4134): 360-366. April 19, 1974.

586. Haggard, Jerry L. "The Mining Law of 1872: Reform or Repeal?" NATURAL RESOURCES LAWYER 6(1): 98-107. Winter 1973.

587. Haight, G. Winthrop. "Libyan Nationalization of British Petroleum Company Assets." THE INTERNATIONAL LAWYER 6(3): 541-547. July 1972.

588. Halacy, Daniel S., Jr. THE COMING AGE OF SOLAR ENERGY. Harper and Row, New York. 1974.

589. Hall, Gus. THE ENERGY RIP-OFF: CAUSE AND CURE. International Publishers, New York. 1974.

590. Hall, Robert E. Lee. "Domestic Coal Vs. Foreign Residual Oil." NATURAL RESOURCES LAWYER 3: 266-270. May 1970.

591. Halvorsen, R. SIERRA CLUB CONFERENCE ON POWER AND PUBLIC POLICY. Public Resources, Inc., Burlington, Vermont. 1972.

592. Hamilton, Richard E. "A Marketing Board to Regulate Exports of Natural Gas?" CANADIAN PUBLIC ADMINISTRATION 16(1): 83-95. Spring 1973.

593. Hamilton, Richard E. "Canada's 'Exportable Surplus' Natural Gas Policy: A Theoretical Analysis." LAND ECONOMICS 49(3): 251-259. August 1973.

594. Hammerman, Howard M. "Toward An Organizational Basis for Human Ecology." Paper Presented at the 1974 Meeting of the American Sociological Association, Montreal, Canada. August 1974.

595. Hammond, A.L. "Energy Options: Challenge for the Future." SCIENCE 178: 875-876. September 8, 1972.

596. Hammond, Allan L. "Conservation of Energy: The Potential for More Efficient Use." SCIENCE 178: 1079-1081. December 8, 1972.

597. Hammond, Allen L. "Energy Needs: Projected Demands and How to Reduce Them." SCIENCE 178: 1186-1188. December 15, 1972.

598. Hammond, Allen L. "Energy and the Future: Research Priorities and National Policy." SCIENCE 179: 164-166. January 12, 1973.

599. Hammond, Allen L. "Energy Conservation." PROCEEDINGS OF THE ACADEMY OF POLITICAL SCIENCE 31(2): 53-62. December 1973.

600. Hammond, Allen L. "Individual Self-Sufficiency in Energy." SCIENCE 184(4134): 278-283. April 19, 1974.

601. Hammond, Allen L. "A Timetable for Expanded Energy Availability." SCIENCE 184(4134): 367-370. April 19, 1974.

602. Hammond, Allen L., Metz, William, and Maugh, Thomas, II. ENERGY AND THE FUTURE. American Association for the Advancement of Science, Washington, D.C. 1973.

603. Hanessian, John, Jr. and Johnson, Jean. "Research Priorities and Global Energy Policy Issues: A Suggested Analytical Framework." Paper Presented at the Conference on Energy Policies and the International Systems, Santa Barbara, California. December 1973.

604. Hanke, Steve H. and Boland, John J. "Thermal Discharges and Policy Alternatives." WATER RESOURCES BULLETIN 8(3): June 1972.

605. Hannon, Bruce. SYSTEM ENERGY AND RECYCLING: A STUDY OF THE BEVERAGE INDUSTRY. CAC Document Number 23. Center for Advanced Computation, University of Illinois, Urbana. January 1972.

606. Hannon, Bruce. "An Energy Standard of Value." ANNALS OF THE AMERICAN ACADEMY OF POLITICAL AND SOCIAL SCIENCE 410( ): 139-153. November 1973.

607. Hannon, Bruce. OPTIONS FOR ENERGY CONSERVATION. Document Number 79. Center for Advanced Computation, University of Illinois, Urbana. 1973.

608. Hansen, Joachim. "Movements in the International Oil Business." AUSSEN POLITIK 23(3): 324-334. 1972.

609. Hardesty, C. Howard, Jr. "Coal and the Energy Crisis." WEST VIRGINIA LAW REVIEW 76(3): 257-266. April 1974.

610. Hardesty, C. Howard, Jr. "Critical Path to Adequate Supplies of Energy." Address May 18, 1972. VITAL SPEECHES: July 1, 1972.

611. Hardesty, C. Howard, Jr. " Energy vs. Ecology." VITAL SPEECHES: March 1, 1973.

612. Hardesty, C. Howard, Jr. "Domestic Energy Supply and the Environment." THE CONFERENCE BOARD RECORD 11(5): 26-28. May 1974.

613. Hardin, Garrett. "The Fifth Option." ECOLOGY TODAY 1( ): 22-23, 46. May 1971.

614. Hardt, John P. "West Siberia: The Quest for Energy." PROBLEMS OF COMMUNISM 22(3): 25-36. May-June 1973.

615. Hargreaves, Reginald. "The Vital Fluid." MILITARY REVIEW 50: 63-69. July 1970.

616. Harris, Marvin. CULTURAL ENERGY. (Forthcoming).

617. Harris, Shearon. "Will Rural Electric Co-ops be Able to Compete?" MANAGEMENT QUARTERLY 13(1): 2-7. Spring 1972.

618. Hartman, R.M. "Environmental Reports and New Power Plant Schedules." PUBLIC UTILITIES FORTNIGHTLY 88( ): 13-22. July 22, 1971.

619. Hartshorn, J.E. POLITICS AND WORLD OIL ECONOMICS: AN ACCOUNT OF THE INTERNATIONAL OIL INDUSTRY IN ITS POLITICAL ENVIRONMENT. Praeger, New York. 1967.

620. Hartshorn, J.E. "Oil Diplomacy: The New Approach." THE WORLD TODAY 29(7): 281-290. July 1973.

621. Hasson, J.A. THE ECONOMICS OF NUCLEAR POWER. Longmans, London. 1965.

622. Hauser, L.G. and Potter, R.F. "The Future of Fossil Fuels." POWER ENGINEERING 75: 40-43. May 1971.

623. Hausz, W., Leeth, G. and Meyer, C. "Eco-Energy Studies at TEMPO." JOURNAL OF PETROLEUM TECHNOLOGY 17(198): 942-945. November 1972.

624. Haveman, Robert. EFFICIENCY AND EQUITY IN NATURAL RESOURCE AND ENVIRONMENTAL POLICY. IES Report Number 18. Institute for Environmental Studies, University of Wisconsin, Madison. 1974.

625. Havemann, Joel. "Why Energy Bill Failed." NATIONAL JOURNAL REPORTS 6(1): 24- . January 5, 1974.

626. Hawkins, Clark A. "Projecting Future Gas Supply." PUBLIC UTILITIES FORTNIGHTLY 84: 39-43. August 14, 1969.

627. Hays, Richard. "Executives in an Energy Crisis." DUN'S 102(6): 104-111. December 1973.

628. Heberlein, Thomas A. "Conservation Information, The Energy Crisis and Electricity Consumption In An Apartment Complex." Department of Rural Sociology, University of Wisconsin, Madison. June 1974.

629. Heilbroner, Robert L. AN INQUIRY INTO THE HUMAN PROSPECT. W.W. Norton, New York. 1974.

630. Heller, Charles A. THE WORLD PETROLEUM INDUSTRY AND ITS IMPACT ON MIDCONTINENT OIL AND GAS ECONOMICS. State Geological Survey of Kansas, Lawrence. 1969.

631. Hellman, Hal. ENERGY IN THE WORLD OF THE FUTURE. M. Evans and Company, New York. 1973.

632. Henry, John P., Jr. and Schmidt, Richard A. "Coal: Still Old Reliable?" ANNALS OF THE AMERICAN ACADEMY OF POLITICAL AND SOCIAL SCIENCE 410( ): 35-51. November 1973.

633. Herendeen, R.A. AN ENERGY INPUT-OUTPUT MATRIX FOR THE UNITED STATES, 1963: USER'S GUIDE. University of Illinois Press, Urbana, Illinois. 1973.

634. Herfindahl, Henry. WATER POWER IN ALASKA: A BIBLIOGRAPHY. Alaska Power Administration, U.S. Department of the Interior, Juneau. 1965.

635. Herfindahl, Orris C. and Kneese, Allen V. ECONOMIC THEORY OF NATURAL RESOURCES. Charles E. Merril Publishing Company, Columbus, Ohio. 1974.

636. Herman, Stephen A. "The Energy Shortage: Some Legal Implications." THE CONFERENCE BOARD RECORD 11(4): 13-17. April 1974.

637. Hess, W.N. "New Horizons in Resource Development: The Role of Nuclear Explosions." GEOGRAPHICAL REVIEW 52: 1-24. January 1972.

638. Hill, George R. "Can Fossil Fuel Be Cleaned Up?" In ENERGY AND THE ENVIRONMENT: A COLLISION OF CRISES, Irwin Goodwin (ed.). Publishing Sciences Group, Acton, Massachusetts. 1973.

639. Hirst, Eric. ELECTRIC UTILITY ADVERTISING AND THE ENVIRONMENT. Oak Ridge National Laboratory, Oak Ridge, Tennessee. 1972.

640. Hirst, Eric. ENERGY CONSUMPTION FOR TRANSPORTATION IN THE U.S. Oak Ridge National Laboratory, Oak Ridge, Tennessee. 1972.

641. Hirst, Eric. "Energy vs. Environment: The Coming Struggle." LIVING WILDERNESS: Winter 1972.

642. Hirst, Eric. "The Energy Cost of Pollution Control." ENVIRONMENT 15(8): 37-44. October 1973.

643. Hirst, Eric. "Transportation Energy Use and Conservation Potential." SCIENCE AND PUBLIC AFFAIRS: BULLETIN OF THE ATOMIC SCIENTISTS 29(9): 36-42. November 1973.

644. Hirst, Eric. "Food-Related Energy Requirements." SCIENCE 184(4133): 134-138. April 12, 1974.

645. Hirst, Eric. "Transportation Energy Conservation: Opportunities and Policy Issues." TRANSPORTATION JOURNAL 13(3): 42-52. Spring 1974.

646. Hirst, Eric and Healy, T. "Electric Energy Requirements for Environmental Protection." Paper Presented to the conference on Energy: Demand, Conservation, and Institutional Problems, M.I.T., Cambridge, Massachusetts. February 1973.

647. Hirst, Eric and Moyers, John C. "Efficiency of Energy Use in the United States." SCIENCE 179: 1299-1304. March 30, 1973.

648. Hittman Associates, Inc. ELECTRICAL POWER SUPPLY AND DEMAND FORECASTS FOR THE UNITED STATES THROUGH 2050. U.S. Government Printing Office, Washington, D.C. 1971.

649. Hittman Associates, Inc. STUDY OF THE FUTURE SUPPLY OF LOW SULFUR OIL FOR ELECTRIC UTILITIES. U.S. Government Printing Office, Washington, D.C. 1971.

650. Hittman Associates. STUDY OF THE FUTURE SUPPLY OF NATURAL GAS FOR ELECTRIC UTILITIES. U.S. Government Printing Office, Washington, D.C. 1971.

651. Hittman Associates, Inc. SURVEY OF NUCLEAR POWER SUPPLY PROSPECTS. National Technical Information Service, Springfield, Virginia. 1972.

652. Hobart, Lawrence. "Power Issues Face New Congress." PUBLIC POWER. January/February 1971.

653. Hobson, J.E. THE ECONOMICS OF SOLAR ENERGY. From World Symposium on Applied Solar Energy. Phoenix, Arizona. 1955.

654. Hodgkins, J.A. SOVIET POWER: ENERGY RESOURCES, PRODUCTION, AND POTENTIAL. Prentice-Hall, Englewood Cliffs, New Jersey. 1961.

655. Hogerton, John F. "The Outlook for Electric Power." PLANNING: February 1973.

656. Hogerton, John F., Geller, Leonard and Gerber, Abraham. THE OUTLOOK FOR URANIUM: A SURVEY OF THE U.S. URANIUM MARKET. S.M. Stoller Associates, New York. 1965.

657. Holcomb, Robert W. "Oil in the Ecosystem." SCIENCE 166: 204-206. October 10, 1969.

658. Holcomb, Robert W. "Power and Fuel Resources." CURRENT HISTORY 58: 330-336, 365. June 1970.

659. Holden, Constance. "Energy: Shortages Loom, but Conservation Lags." SCIENCE 180(4091): 1155-1158. June 15, 1973.

660. Holdren, John P. and Herrera, Phillip. ENERGY. Sierra Club Books, New York. 1971.

661. Holling, Crawford S. "Stability in Ecological and Social Systems." In DIVERSITY AND STABILITY IN ECOLOGICAL SYSTEMS, Brookhaven Symposia in Biology, Number 22, 1969. (pp. 128-141).

662. Horton, Jack K. "Nuclear Power -- Promise or Problem?" EDISON ELECTRIC INSTITUTE BULLETIN 37: 207-212. June-July 1969.

663. Hottel, Hoyt C. and Howard, Jack B. NEW ENERGY TECHNOLOGY: SOME FACTS AND ASSESSMENTS. M.I.T. Press, Cambridge, Massachusetts. 1971.

664. Hottel, Hoyt C. and Howard, Jack B. "An Agenda for Energy." TECHNOLOGY REVIEW: January 1972.

665. Hoult, David P. (ed.). OIL ON THE SEA. Symposium, Cambridge, Massachusetts, May 1969. Plenum, New York. 1969.

666. Housago, David. "Iran in the Ascendant." THE ROUND TABLE 248: 497-508. October 1972.

667. Houthakker, Hendrik S. and Taylor, Lester D. CONSUMER DEMAND IN THE UNITED STATES. M.I.T. Press, Cambridge, Massachusetts. 1970.

668. Hovanesian, Archie, Jr. "Post TORREY CANYON: Toward a New Solution to the Problem of Traumatic Oil Spillage." CONNECTICUT LAW REVIEW 2( ): 632-647. Spring 1970.

669. Howard, Peter R. "Electrical Transmission of Energy: Current Trends." ENERGY POLICY 1(2): 154-160. September 1973.

670. Hub, K. et al. A STUDY OF SOCIAL COSTS FOR ALTERNATE MEANS OF ELECTRIC POWER GENERATION FOR 1980 and 1990. Argonne National Laboratory, Argonne, Illinois. 1973.

671. Hubbert, M. King. ENERGY RESOURCES. National Academy of Sciences, Washington, D.C. 1962.

672. Hubbert, M. King. "Energy Resources." In RESOURCES AND MAN: A STUDY AND RECOMMENDATIONS, National Research Council, National Academy of Sciences. W.H. Freeman, San Francisco. 1969. (pp. 157-242).

673. Hubbert, M. King. "The Energy Resources of the Earth." SCIENTIFIC AMERICAN 225: 60-70. September 1971.

674. Hunter, Robert E. THE SOVIET DILEMMA IN THE MIDDLE EAST, PART II: OIL AND THE PERSIAN GULF. Adelphi Paper Number 60. Institute for Strategic Studies, London 1969.

675. Hunter, Robert E. THE ENERGY 'CRISIS' AND U.S. FOREIGN POLICY. Headline Series Number 216. Foreign Policy Association, New York. 1973.

676. Hurlbut, Cornelius S. MINERALS AND MAN. Random House, New York. 1968.

677. Hurstfield, J. THE CONTROL OF RAW MATERIALS. Her Majesty's Stationery Office, London. 1953.

678. Ikard, Frank N. "Criticism, Policy and Reality: A National Energy Policy." Address May 28, 1971. VITAL SPEECHES: August 1, 1971.

679. Ikard, Frank N. "Energy and Economics." Address, July 24, 1972. VITAL SPEECHES: September 15, 1972.

680. Ikard, Frank N. "The Oil Industry and the Energy-Environment Problem." In ENERGY AND THE ENVIRONMENT: A COLLISION OF CRISES, Irwin Goodwin (ed.). Publishing Sciences Group, Acton, Massachusetts. 1973.

681. Illich, Ivan. ENERGY AND EQUITY. Harper and Row, New York. 1974.

682. IMPACT OF SCIENCE ON SOCIETY (eds.). "Man's Use of Energy." IMPACT OF SCIENCE ON SOCIETY 1. October/December 1950. (Whole Issue).

683. Imsland, Donald O. "Futurizing a Power Company." THE FUTURIST 7(4): 175-176. August 1973.

684. INDIANA LAW JOURNAL. "Thermal Electric Power and Water Pollution: A Siting Approach." INDIANA LAW JOURNAL 46( ): 61- . 1970.

685. INDIANA LAW JOURNAL. "The Four Corners Power Complex: Pollution on the Reservations." INDIANA LAW JOURNAL 47( ): 704- . 1971.

686. INDIANA LAW JOURNAL. "Natural Gas and the Federal Power Commission." INDIANA LAW JOURNAL 47(4): 742-754. Summer 1972.

687. Inglis, David Rittenhouse. "Nuclear Energy and the Malthusian Dilemma." BULLETIN OF THE ATOMIC SCIENTISTS 27: 14-18. February 1971.

688. Inglis, K.A. (ed.). ENERGY -- FROM SURPLUS TO SCARCITY? Halsted Press, New York. 1974.

689. Institute of Electrical and Electronic Engineers. THE GREAT ENVIRONMENTAL DEBATE AND THE POWER INDUSTRY. Institute of Electrical and Electronic Engineers, New York. 1970.

690. Institute on Economics of the Gas Industry, Dallas, 1962. ECONOMICS OF THE GAS INDUSTRY: COMPETITION, CONSERVATION, CONTROL, INVESTMENT, FINANCING, MARKETING, GAS RESERVES, OCCURRENCE, PRODUCTION. M. Bender, Albany. 1962.

691. International Atomic Energy Agency. LIST OF BIBLIOGRAPHIES ON NUCLEAR ENERGY. International Atomic Energy Agency, Vienna. 1970.

692. International Atomic Energy Agency. "Nuclear Power and the Environment." INTERNATIONAL ATOMIC ENERGY AGENCY BULLETIN 12(5): 9-21. 1970.

693. International Institute for Environmental Affairs. WORLD ENERGY, THE ENVIRONMENT AND POLITICAL ACTION. International Institute for Environmental Affairs, New York. 1973.

694. Interstate Oil Compact Commission. A STUDY OF CONSERVATION OF OIL AND GAS IN THE UNITED STATES. Interstate Oil Compact Commission, Tulsa, Oklahoma. 1964.

695. Interstate Oil Compact Commission. Engineering Committee. PRINCIPLES OF PETROLEUM CONSERVATION: A REPORT. Interstate Oil Compact Commission, Oklahoma City. 1969.

696. InterTechnology Corporation. THE U.S. ENERGY PROBLEM. InterTechnology Corporation, Warrenton, Virginia. 1971.

697. Ion, D.C. "The Significance of World Petroleum Reserves." Paper Presented at the Seventh World Petroleum Congress, Mexico City. April 1967.

698. Irwin, J.N., 2nd. "International Implications of the Energy Situation." Statement April 10, 1972. U.S. DEPARTMENT OF STATE BULLETIN 66: 626-631. May 1, 1972.

699. Ise, John. THE UNITED STATES OIL POLICY. Yale University Press, New Haven, Connecticut. 1926.

700. Ismael, Salem K. THE CORRELATION BETWEEN ENERGY CONSUMPTION AND GNP. Organization of Petroleum Exporting Countries, Vienna. 1968.

701. Issawi, Charles. "Oil and Middle East Politics." PROCEEDINGS OF THE ACADEMY OF POLITICAL SCIENCE 31(2): 111-122. December 1973.

702. Issawi, Charles. "Consequences of the Oil Squeeze." INTERNATIONAL PERSPECTIVES: 9-12. March-April 1974.

703. Issawi, Charles P. and Yeganeh, Mohammed. THE ECONOMICS OF MIDDLE EASTERN OIL. Praeger, New York. 1963.

704. Iulo, William. "Supply and Demand for Energy: Largely Domestic." TRANSPORTATION JOURNAL 13(3): Spring 1974.

705. Jackson, Raymond. "Regulation and Electric Utility Rate Levels." LAND ECONOMICS 45: 372-376. August 1969.

706. Jacobsen, A. PETROLEUM: RECOMMENDATIONS FOR A NATIONAL OIL POLICY. U.S. Government Printing Office, Washington, D.C. 1947.

707. James, J.R., Scott, Sheila F. and Willatts, E.C. "Land Use and the Changing Power Industry in England and Wales." GEOGRAPHICAL JOURNAL 127: 286-309. September 1961.

708. James, L. Douglas and Lee, Robert R. ECONOMICS OF WATER RESOURCES PLANNING. McGraw-Hill, New York. 1971.

709. Jameson, Minor S., Jr. "Policy Criteria for Petroleum." In PERSPECTIVES ON CONSERVATION: ESSAYS ON AMERICA'S NATURAL RESOURCES, Henry Jarrett (ed.). Published for Resources for the Future by The Johns Hopkins Press, Baltimore. 1958. (pp. 191-195).

710. Jefferson, Edward G. "Energy Management in Industry." THE CONFERENCE BOARD RECORD 11(5): 37-45. May 1974.

711. Jensen, James T. and Stauffer, Thomas R. IMPLICATIONS OF NATURAL GAS CONSUMPTION PATTERNS FOR THE IMPLEMENTATION OF END-USE PRIORITY PROGRAMS. Arthur D. Little, Inc., Cambridge, Massachusetts. 1972.

712. Jensen, Walter G.W. ENERGY AND THE ECONOMY OF NATIONS. G.T. Foulis, Henley-on-Thames, England. 1970.

713. Jimison, John and Lieberman, Joseph. A REVIEW OF THE ENERGY POLICY ACTIVITIES OF THE 92D CONGRESS. U.S. Government Printing Office, Washington, D.C. 1973.

714. Johnson, Arthur M. PETROLEUM PIPELINES AND PUBLIC POLICY: 1906-1959. Harvard University Press, Cambridge, Massachusetts. 1967.

715. Johnson, Sue and Burdge, Rabel. "A Methodology for Using Diachronic Studies to Predict the Social Impact of Resource Development." Center for Developmental Change, University of Kentucky, Lexington. 1974.

716. Johnson, Sue and Randall, Alan. "Some Social and Economic Considerations in Coal Conversion Technology." Paper Prepared for a Workshop on Research Needs Related to Water for Energy, Sponsored by the Office of Water Research and Technology, Indianapolis, Indiana. October 1974.

717. Johnson, Thomas G. "The Hearing Under Section 7 of the Natural Gas Act -- What Now?" NATURAL RESOURCES LAWYER 2: 200-207. July 1969.

-50-

718. Johnson, William A. "Solving the Energy Problem: An Analysis of the Issues." THE CONFERENCE BOARD RECORD 11(4): 24-26. April 1974.

719. Johnston, Warren E. "Natural Resources Economics Education in the Land Grant 'West'." In ECONOMICS OF NATURAL RESOURCE DEVELOPMENT IN THE WEST: CURRENT PROBLEMS IN NATURAL RESOURCE USE, E. Boyd Wennergren (ed.). Committee on the Economics of Natural Resources Development, Western Agricultural Economics Research Council, Utah State University, Logan. 1973. (pp. 166-179).

720. Jones, M.V. THE IMPACT ASSESSMENT SCENARIO, A PLANNING TOOL FOR MEETING THE NATION'S ENERGY NEEDS. M72-56. Mitre Corporation, McLean, Virginia. April 1972.

721. Jopling, David, Gage, Stephen and Schoeman, Milton. "Forecasting Public Resistance to Technology: The Example of Nuclear Power Reactor Siting." In A GUIDE TO PRACTICAL TECHNOLOGICAL FORECASTING, James R. Bright and Milton E.F. Schoeman (eds.). Prentice-Hall, Englewood Cliffs, New Jersey. 1973. (pp. 53-66).

722. Jordan, Walter H. "Nuclear Energy: Benefits Versus Risks." PHYSICS TODAY 23: 32-38. May 1970.

723. Josephy, A. M., Jr. "The Murder of the Southwest." AUDUBON MAGAZINE ( ): July 1971.

724. Josephy, A. M., Jr. "Agony of the Northern Plains." AUDUBON MAGAZINE ( ): July 1973.

725. Joskow, Paul L. "Inflation and Environmental Concern: Structural Change in the Process of Public Utility Price Regulation." March 1974.

726. JOURNAL OF NUTRITION EDUCATION (eds.). "Energy Crisis Today, Food Crisis Tomorrow? Editorial." JOURNAL OF NUTRITION EDUCATION 6(1): 4-7. January-March 1974.

727. Journey, Drexel D. "Power Plant Siting, a Road Map of the Problem." NOTRE DAME LAW 48( ): 273- . December 1972.

728. Kafoglis, Milton and Keig, Norman. "New Policies of the Federal Power Commission." LAND ECONOMICS 45: 385-391. November 1969.

729. Kahn, Alfred E. "The Combined Effects of Prorationing, the Depletion Allowance and Import Quotas on the Cost of Producing Crude Oil in the United States." NATURAL RESOURCES JOURNAL 10: 53-61. January 1970.

730. Kahn, Alfred E. THE ECONOMICS OF REGULATION: PRINCIPLES AND INSTITUTIONS. John Wiley and Sons, New York. 1970.

731. Kalter, Robert J. et al. CRITERIA FOR FEDERAL EVALUATION OF RESOURCE INVESTMENTS. Water Resources and Marine Sciences Center, Cornell University, Ithaca, New York. 1969.

732. Kaplan, Marcos. "La Politica de Petroleo (1907-1955)." FORO INTERNACIONAL 14(1): 85-105. July-September 1973.

733. Karp, Richard. "Atomic Power Abuse: The Marginal Nuclear Utilities." WASHINGTON MONTHLY 2: 15-22. July 1970.

734. Kash, Don E. ENERGY UNDER THE OCEANS. University of Oklahoma Press, Norman. 1973.

735. Kash, Don E. and White, Irvin L. "A Preliminary Proposal for a Technology Assessment of Offshore Oil Operations." The Science and Public Policy Program, University of Oklahoma, Norman. November 1970.

736. Kasper, Raphael G. (ed.). TECHNOLOGY ASSESSMENT: UNDERSTANDING THE THE SOCIAL CONSEQUENCES OF TECHNOLOGICAL APPLICATIONS. Praeger, New York. 1972.

737. Katz, M. "Criticism, Policy and Reality: A National Energy Policy." VITAL SPEECHES 37( ): 625-528. August 1, 1971.

738. Kaufman, Alvin. "Beauty and the Beast: The Siting Dilemma in New York State." ENERGY POLICY 1(3): 243-253. December 1973.

739. Kaufman, Irving R. "Power for the People -- And by the People: Utilities, the Environment and the Public Interest." ADMINISTRATIVE LAW REVIEW 24(1): 3-14. Winter 1972.

740. Keenan, Boyd R. "Energy Crisis and Its Meaning for American Culture." CHRISTIAN CENTURY 90: 756-759. July 18, 1973.

741. Keener, Kenneth C. "Federal Water Pollution Legislation and Regulations With Particular Reference to the Oil Industry." NATURAL RESOURCES LAWYER 2( ): 484-504. July 1971.

742. Kelly, Peter M. "The Power Crisis and the Counties." AMERICAN COUNTY: March 1973.

743. Kemp, W.B. "The Flow of Energy in a Hunting Society." SCIENTIFIC AMERICAN 224: 104-115. September 1971.

744. Kennedy, Michael. "Energy: End of an Era?" COMMONWEALTH 99(21): 527-529. March 1, 1974.

745. Kennedy, W.F. "Nuclear Electrical Power and the Environment -- New Regulatory Structures and Procedures." ATOMIC ENERGY LAW JOURNAL 13( ): 293- . 1972.

746. Khazzoom, J. Daniel. "The FPC Staff's Econometric Model of Natural Gas Supply in the United States." BELL JOURNAL OF ECONOMICS AND MANAGEMENT SCIENCE 2(1): 51-93. Fall 1971.

747. Kiley, Edward W. "Britain's Electricity Crisis." MANAGEMENT QUARTERLY 13(1): 8-13. Spring 1972.

748. Kindred, Hugh M. and Schwartz, Warren F. "American Regulation of Oil Imports: Law Policy and Institutional Responsibility." JOURNAL OF WORLD TRADE LAW 5( ): 267-302. May-June 1971.

749. King, Kerryn. "Petroleum Energy and the Environment." NATURAL RESOURCES LAWYER 4( ): 780-789. November 1971.

750. King, Peter. "Viewpoint: Oil: The New International Lever." THE BUSINESS QUARTERLY 39(2): 5-20. Summer 1974.

751. Kinney, Gene T. "The Environmental Craze: Will it Strangle Energy?" OIL AND GAS JOURNAL 69: 23-28. March 15, 1971.

752. Kjolberg, Anders. "Political Barriers to Supplying Gas and Oil." INTERNASJONAL POLITIKK (3): 515-532. July-September 1973.

753. Klaff, Jerome L. "Power Policies." PUBLIC UTILITIES FORTNIGHTLY: September 13, 1973.

754. Klineberg, Otto. SOCIAL IMPLICATIONS OF THE PEACEFUL USES OF NUCLEAR ENERGY. UNESCO, Paris. 1964.

755. Klingstedt, John P. "Effects of Full Costing in the Petroleum Industry." FINANCIAL ANALYSTS JOURNAL 26: 79-86. September-October 1970.

756. Kloman, Erasmus H. "Public Participation in Technology Assessment." PUBLIC ADMINISTRATION REVIEW: 52-61. January/February 1974.

757. Knetsch, Jack L. et al. FEDERAL NATURAL RESOURCES DEVELOPMENT: BASIC ISSUES IN BENEFIT AND COST MEASUREMENT. Natural Resources Policy Center, George Washington University, Washington, D.C. n.d.

758. Knight, Andrew (ed.). "The Big League: Petrochemicals: A Survey." ECONOMIST 237: October 3, 1970. (Whole Issue).

759. Knoll, Erwin. "The Oil Lobby is not Depleted." NEW YORK TIMES MAGAZINE: 26-27, 103-109. March 8, 1970.

760. Knowles, Ruth. GREATEST GAMBLERS: THE EPIC OF AMERICAN OIL EXPLORATION. McGraw-Hill, New York. 1959.

761. Knox, Susan. THE ENERGY CRISIS SURVIVAL KIT. Manor Books, New York. 1974.

762. Koeller, J.R. "Conservation of Energy." NAVY CIVIL ENGINEER 13( ): 18-19. Winter 1972.

763. Kolin, Horst. "Energy Supplies in Austria." EUROPEAN FREE TRADE ASSOCIATION BULLETIN 15(4): 6-7. May 1974.

764. Kouba, J.H. "Regulating Electric Transmission Lines in California: Insulation from Aesthetic Shock?" HASTINGS LAW JOURNAL 22( ): 587- . 1971.

765. Kraar, Louis. "Oil and Nationalism Mix Beautifully in Indonesia." FORTUNE 88(1): 99-103. July 1973.

766. Kramer, Eugene. "Energy Conservation and Waste Recycling: Taking Advantage of Urban Congestion." SCIENCE AND PUBLIC AFFAIRS - BULLETIN OF THE ATOMIC SCIENTISTS 29(4): 13-18. April 1973.

767. Krieger, James H. "Energy: The Squeeze Begins." CHEMICAL AND ENGINEERING NEWS 50: 20-22, 24-28, 33-37. November 13, 1972.

768. Krier, J.E. "The Pollution Problem and Legal Institutions: A Conceptual Overview." UCLA LAW REVIEW 18( ): 429- . Fall 1971.

769. Kuhnke, Hans-Helmut. "The Long-Term Energy Supply." INTERECONOMICS (2): February 1970.

770. Kulcinski, Gerald L. "Fusion Power: An Assessment of its Potential Impact." ENERGY POLICY 2(2): 104-125. June 1974.

771. Kvinnsland, Ole-Jacob. "Norwegian Oil -- Economic and Political Problems in a European Context." INTERNASJONAL POLITIKK (4): 581-591. 1972.

772. Kwee, S.L. and Mullander, J.S.R. (eds.). GROWING AGAINST OURSELVES: THE ENERGY-ENVIRONMENT TANGLE. Heath-Lexington, Lexington, Massachusetts. 1972.

773. Lacewell, Ronald D. "Linear Programming Model for Resource Planning and Energy Considerations." In ECONOMICS OF NATURAL RESOURCE DEVELOPMENT IN THE WEST: CURRENT PROBLEMS IN NATURAL RESOURCE USE, E. Boyd Wennergren (ed.). Committee on the Economics of Natural Resources Development, Western Agricultural Economics Research Council, Utah State University, Logan. 1973. (pp. 50-55).

774. Laird, W.M. "Energy: A Scarcity of All Kinds." Address March 5, 1970. VITAL SPEECHES: April 15, 1970.

775. LaLonde, Bernard J. "The Energy Gap and Distribution Strategy." BULLETIN OF BUSINESS RESEARCH 49(4): 1-3. April 1974.

776. Lambin, Jean-Jacques. "Is Gasoline Advertising Justified?" THE JOURNAL OF BUSINESS 45(4): 585-619. October 1972.

777. Landsberg, Hans H. NATIONAL RESOURCES FOR U.S. GROWTH -- A LOOK AHEAD TO THE YEAR 2000. Johns Hopkins, Baltimore. 1964.

778. Landsberg, Hans H. FACTORS IN THE LONG RANGE COMPETITIVE SETTING OF OIL SHALE. Resources for the Future, Washington, D.C. 1965.

779. Landsberg, Hans H. "Materials Resources 1990 and Beyond: What Will We Have Left to Work With?" Paper Presented at the Annual Meeting of the American Association for the Advancement of Science, San Francisco. February 1974.

780. Landsberg, Hans H. "Low-Cost, Abundant Energy: Paradise Lost?" SCIENCE 184(4134): 247-254. April 19, 1974.

781. Landsberg, Hans H., Fischman, Leonard L. and Fisher, Joseph L. RESOURCES IN AMERICA'S FUTURE: PATTERNS OF REQUIREMENTS AND AVAILABILITIES, 1960-2000. Published for Resources for the Future by the Johns Hopkins Press, Baltimore. 1963.

782. Landsberg, Hans H. and Schurr, Sam H. ENERGY IN THE UNITED STATES: SOURCES, USES, AND POLICY ISSUES. Random House, New York. 1968.

783. Langdon, Jim C. "The Energy Crisis and the Producer States." NATURAL RESOURCES LAWYER 6(4): 485-494. Fall 1973.

784. Lapp, Ralph E. A CITIZEN'S GUIDE TO NUCLEAR POWER. New Republic, Washington. 1971.

785. Lapp, Ralph E. "Greater R&D Effort Required to Meet Future Energy Crisis." PUBLIC POWER 30( ): 12-15. January-February 1972.

786. Lapp, Ralph E. THE LOGARITHMIC CENTURY. Prentice-Hall, Englewood Cliffs, New Jersey. 1973.

787. Lapp, Ralph E. "The Ultimate Blackmail." NEW YORK TIMES MAGAZINE: February 4, 1973.

788. Large, David B. (ed.). HIDDEN WASTE: POTENTIALS FOR ENERGY CONSERVATION. Conservation Foundation, Washington, D.C. 1973.

789. Larson, C.E. "Present State and Future Outlook of Nuclear Power Generation in the U.S." ATOMIC ENERGY LAW JOURNAL 12: 274. Fall 19

790. Laster, L.L. ATMOSPHERIC EMISSIONS FROM PETROLEUM REFINING INDUSTRY. National Technical Information Service, Springfield, Virginia. 1973.

791. Latus, Mark. "Preservation and the Energy Crisis." HISTORIC PRESERVATION: April/June 1973.

792. Laughton, M.A. "Energy: Demand, Conservation and Institutional Problems." ENERGY POLICY 1(1): 71- . June 1973.

793. Lave, Lester B. and Freeburg, Linnea C. "Health Costs to the Consumer Per Kilowatt." In ENERGY, THE ENVIRONMENT, AND HUMAN HEALTH, Asher J. Finkel (ed.). Publishing Sciences Group, Acton, Massachusetts. 1973.

794. LAW AND SOCIAL ORDER. "Power Plant and Transmission Line Siting: Improving Arizona's Legislative Approach." LAW AND SOCIAL ORDER 1973( ): 519- . 1973.

795. Lawrence, Robert. "Bibliographical Essay on Energy Policy." POLICY STUDIES JOURNAL 2(2): 141- . Winter 1973.

796. Lawrence, Robert M. and Wengert, Norman I. "Preface -- The Energy Crisis: Reality or Myth." ANNALS OF THE AMERICAN ACADEMY OF POLITICAL AND SOCIAL SCIENCE 410( ): ix- . November 1973.

797. Laxer, J. THE ENERGY POKER GAME: THE POLITICS OF THE CONTINENTAL RESOURCES DEAL. New Press, Toronto. 1970.

798. Laycock, George. "Kiss the North Slope Good-by? It's Oil Country Now!" AUDUBON 72: 58-75. September 1970.

799. Leach, Gerald. "The Impact of the Motor Car on Oil Reserves." ENERGY POLICY 1(3): 195-207. December 1973.

800. Leachman, Robert B. "Final Report: Political and Scientific Effectiveness in Nuclear Materials Control." Report to the National Science Foundation on Grant Number GI-9. May 1972.

801. Leachman, Robert B. and Althoff, Philip. PREVENTING NUCLEAR THEFT: GUIDELINES FOR INDUSTRY AND GOVERNMENT. Praeger, New York. 1972.

802. League of Women Voters. Education Fund. "Fact Sheets on Energy." League of Women Voters, Washington, D.C. 1974.

803. Leavitt, Helen. "Alternative to Automobilia." NATION'S CITIES 10(4): 24-26. April 1972.

804. LeBoeuf, R.J., Jr. "An Industry Appraisal of Federal Regulation of Electric Utilities Under the Federal Power Act." GEORGE WASHINGTON LAW REVIEW 14( ): 174-193. December 1945.

805. Lee, Richard B. "Energy Flow in Hunter-Gatherer Society." Paper Presented at the Annual Meeting of the American Association for the Advancement of Science, San Francisco. February 1974.

806. Leeston, Alfred M. et al. THE DYNAMIC NATURAL GAS INDUSTRY. University of Oklahoma Press, Norman. 1963.

807. Leholm, Arlen, Leistritz, F. Larry and Hertsgaard, Thor. "Local Impacts of Energy Resources Development in the Northern Great Plains." Department of Agricultural Economics, North Dakota State University, Fargo. April 1974.

808. Lenczowski, George. "Multinational Oil Companies: A Factor in Middle East International Relations." CALIFORNIA MANAGEMENT REVIEW 8: 38-44. Winter 1970.

809. Lenczowski, George. "Probing the Arab Motivations." INTERNATIONAL PERSPECTIVES: 3-8. March-April 1974.

810. Lessing, Lawrence. "New Ways to More Power With Less Pollution." FORTUNE 82: 78-81, 131-132, 136. November 1970.

811. Lessing, Lawrence. "The Coming Hydrogen Economy." FORTUNE 86(5): 138-147. November 1972.

812. Levy, Walter J. "The Dilemma Posed in Achieving Energy Security." THE CONFERENCE BOARD RECORD 9(7): 20. July 1972.

813. Levy, Walter J. "World Oil Cooperation or International Chaos." FOREIGN AFFAIRS 52(4): 690-713. July 1974.

814. Lewis, Richard S. THE NUCLEAR-POWER REBELLION: CITIZENS VS. THE ATOMIC INDUSTRIAL ESTABLISHMENT. The Viking Press, Inc., New York. 1972.

815. Lewis, Richard S. and Spinrad, Bernard I. (eds.). THE ENERGY CRISIS: A SCIENCE AND PUBLIC AFFAIRS BOOK. Educational Foundation for Nuclear Science, Chicago. 1972.

816. Lieberman, Joseph A. and Belter, Walter G. "Waste Management and Environmental Aspects of Nuclear Power." ENVIRONMENTAL SCIENCE AND TECHNOLOGY 1: 466-475. June 1967.

817. Like, Irving. "Multi-Media Confrontation -- The Environmentalists Strategy for a 'No-Win' Agency Proceeding." ATOMIC ENERGY LAW JOURNAL 13( ): 1- . Spring 1971.

818. Limaye, Dilip R. and Sharko, John R. "U.S. Energy Policy Evaluation: Some Analytical Approaches." ENERGY POLICY 2(1): 3-17. March 1974.

819. Lincoln, G.A. "Energy Conservation." SCIENCE 180: 155. April 13, 1973.

820. Lindahl, David M. THE GASOLINE SHORTAGE: A NATIONAL PERSPECTIVE. U.S. Government Printing Office, Washington, D.C. 1973.

821. Linden, Henry R. "Current Trends in U.S. Gas Demand and Supply." PUBLIC UTILITIES FORTNIGHTLY 86: 27-38. July 30, 1970.

822. Lippitt, Henry F., II. "State and Federal Regulatory Agencies -- Conflict or Cooperation?" PUBLIC UTILITIES FORTNIGHTLY 85: 33-38. March 26, 1970.

823. Little, Arthur D., Inc. ENERGY POLICY ISSUES FOR THE UNITED STATES DURING THE SEVENTIES. National Energy Forum, Arlington, Virginia. 1971.

824. Lobel, Martin. "Red, White, Blue & Gold: The Oil Import Quotas." WASHINGTON MONTHLY 2: 8-18. August 1970.

825. Lockwood, Robert M. "The Energy Economy: Energy in Perspective." TEXAS BUSINESS REVIEW: June 1973.

826. Löf, George O.G. "Solar Energy: An Infinite Source of Clean Energy." ANNALS OF THE AMERICAN ACADEMY OF POLITICAL AND SOCIAL SCIENCE 410( ): 52-64. November 1973.

827. Löf, George O.G., Duffie, John A. and Smith, Clayton O. WORLD DISTRIBUTION OF SOLAR RADIATION. Report Number 21. Engineering Experiment Station. University of Wisconsin Press, Madison. 1966.

828. Lollock, Donald L. "Temperature -- Biological Aspects Related to Nuclear Power Plant Siting, Operation, and Other Considerations." THE FORUM 8(2): 381-410. Winter 1972.

829. Loomis, Carol J. "How to Think About Oil-Company Profits." FORTUNE 89(4): 98-103. April 1974.

830. Lovejoy, Wallace F. "Oil Conservation, Producing Capacity, and National Security." NATURAL RESOURCES JOURNAL 10: 62-96. January 1970.

831. Lovejoy, Wallace F. and Homan, Paul T. PROBLEMS OF COST ANALYSIS IN THE PETROLEUM INDUSTRY. Southern Methodist University Press, Dallas. 1964.

832. Lovejoy, Wallace F. and Homan, Paul T. METHODS OF ESTIMATING RESERVES OF CRUDE OIL, NATURAL GAS, AND NATURAL GAS LIQUIDS. Published for Resources for the Future by the Johns Hopkins Press, Baltimore. 1965.

833. Lovejoy, Wallace F. and Homan, Paul T. ECONOMIC ASPECTS OF OIL CONSERVATION REGULATION. The Johns Hopkins Press, Baltimore. 1967.

834. Lovejoy, Wallace F. and Pikl, I. James, Jr. ESSAYS ON PETROLEUM CONSERVATION REGULATION. Southern Methodist University Press, Dallas. 1960.

835. Lovins, A.B. "World Energy Strategies." SCIENCE AND PUBLIC AFFAIRS 30(5): 13-32. May 1974.

836. Lubell, Harold. MIDDLE EAST OIL CRISES AND WESTERN EUROPE'S ENERGY SUPPLIES. Johns Hopkins Press, Baltimore. 1963.

837. Luce, Charles F. "Power for Tomorrow: The Siting Dilemma." RECORD OF THE ASSOCIATION OF THE BAR OF THE CITY OF NEW YORK 25: 13-26. January 1970.

838. Luidblow, Charles E. "The Science of 'Muddling Through'." In POLITICS, POLICY AND NATURAL RESOURCES, Dennis L. Thompson (ed.). MacMillan, New York. 1972.

839. Lustig, Harry. "Our Dwindling Energy Resources." UNESCO COURIER (1): 4-5. January 1974.

840. Luten, Daniel B. "Economic Geography of Energy." SCIENTIFIC AMERICAN: September 1971.

841. Luten, Daniel B. "United States Energy Requirements." In ENERGY, THE ENVIRONMENT, AND HUMAN HEALTH, Asher J. Finkel (ed.). Publishing Sciences Group, Acton, Massachusetts. 1973.

842. McAfee, Jerry. "Canadian Gas and Oil: Anything Left to Export?" THE CONFERENCE BOARD RECORD 10(8): 31-32. August 1973.

843. MacAllan, D.H. "Government and the Oil Industry -- Quo Vadis." THE BUSINESS QUARTERLY 39(2): 56-61. Summer 1974.

844. MacAvoy, Paul W. PRICE FORMATION IN THE NATURAL GAS FIELD: A STUDY OF COMPETITION MONOPSONY, AND REGULATION, Yale University Press, New Haven. 1962.

845. MacAvoy, Paul W. ECONOMIC STRATEGY FOR DEVELOPING NUCLEAR BREEDER REACTORS. MIT Press, Cambridge, Massachusetts. 1969.

846. MacAvoy, Paul W. "The Effectiveness of the Federal Power Commission." BELL JOURNAL OF ECONOMICS AND MANAGEMENT SCIENCE 1: 271-303. Autumn 1970.

847. MacAvoy, Paul W. "Federal Power Commission and Coordination Problem in the Electrical Power Industry." SOUTHERN CALIFORNIA LAW REVIEW 46( ): 661- . 1973.

848. MacAvoy, Paul W. and Pindyck, Robert S. "Alternative Regulatory Policies for Dealing with the Natural Gas Shortage." THE BELL JOURNAL OF ECONOMICS AND MANAGEMENT SCIENCE 4(2): 454-498. Autumn 1973.

849. McCallum, John and Faust, Charles L. "New Frontiers in Energy Storage." BATTELLE RESEARCH OUTLOOK 4(1): 1972.

850. McCaull, Julian. "Energy and the Worker." ENVIRONMENT 16(6): 35-39. July-August 1974.

851. McCloskey, Michael. "Energy Crisis: Views of an Environmentalist." Address, June 15, 1971. VITAL SPEECHES: August 1, 1971.

852. McCloskey, Michael. "The Energy Crisis: The Issues and a Proposed Response." ENVIRONMENTAL AFFAIRS 1(3): 587-605. November 1971.

853. McCluney, Ross. "Electrical Power Generation and the Environment." FLORIDA NATURALIST 45( ): 56-59. February 1972.

854. McCormack, Mike. "We Must Explore New Methods of Energy Conversion." PUBLIC POWER DIRECTORY: January/February 1973.

855. McCracken, Paul W. "The U.S. Fuel and Energy Situation." CONFERENCE BOARD RECORD 8( ): 37-40. March 1971.

856. McCracken, Paul W. "Energy Economics." THE CONFERENCE BOARD RECORD 9(7): 21. July 1972.

857. McDonald, John. "Oil and the Environment: The View From Maine." FORTUNE 83: 84-89, 146-147, 150. April 1971.

858. McDonald, Stephen L. "Distinctive Tax Treatment of Income from Oil and Gas Production." NATURAL RESOURCES JOURNAL 10: 97-112. January 1970.

859. McDonald, Stephen L. PETROLEUM CONSERVATION IN THE UNITED STATES: AN ECONOMIC ANALYSIS. Published for Resources for the Future by the Johns Hopkins Press, Baltimore. 1971.

860. McDonald, Stephen L. "Public Policy and the Future Adequacy of Oil and Gas Supplies." TEXAS BUSINESS REVIEW 46( ): 163-168. August 1972.

861. McDonald, Stephen L. "Energy in the USA After the President's Messages...Likely Effects on Supply and Demand." ENERGY POLICY 1(3): 179-186. December 1973.

862. McGee, Dean A. "Balancing the Demand and Supply of Electricity and Nuclear Fuels." In BALANCING SUPPLY AND DEMAND FOR ENERGY IN THE UNITED STATES, Rocky Mountain Petroleum Institute, University of Denver. 1972.

863. McGee, Dean A. "Assessing the 'Energy Crisis' -- Problems and Prospects." FEDERAL RESERVE BANK OF KANSAS CITY MONTHLY REVIEW: 14-20. 14-20. September-October 1973.

864. McGeorge, Robert L. "Approaches to State Taxation of the Mining Industry NATURAL RESOURCES JOURNAL 10: 156-170. January 1970.

865. McGovern, George. "Power and Public Responsibility." RURAL ELECTRIFICATION: 36, 40-41. February 1971.

866. McGuire, E. Patrick. "Living With Scarcity." THE CONFERENCE BOARD RECORD 11(3): 5-9. March 1974.

867. McHale, John. "World Energy Resources in the Future." FUTURES: September 1968.

868. McKelvey, V.E. "Mineral Resource Estimates and Public Policy." AMERICAN SCIENTIST 60(1): 32-40. January-February 1972.

869. McKelvey, V.E. and Duncan, D.C. "United States and World Resources of Energy." SYMPOSIUM ON FUEL AND ENERGY ECONOMICS (AMERICAN CHEMICAL SOCIETY) 9(2): 1-17. 1965.

870. MacKenzie, James J. "Social Costs of Energy." MASSACHUSETTS AUDUBON NEWSLETTER 10: 3-6. April 1971.

871. McKetta, John J. "U.S. Warned About Energy Chaos." WORLD OIL 174 (7): 76-78. June 1972.

872. McKetta, John J. "The Energy Crisis: On and On and On." CHEMICAL ENGINEERING PROGRESS 69: 51-56. August 1973.

873. McKie, James W. THE REGULATION OF NATURAL GAS. American Enterprise Association, Washington, D.C. 1957.

874. McKie, James W. "The Political Economy of World Petroleum." THE AMERICAN ECONOMIC REVIEW 64(2): 51-57. May 1974.

875. McLane, James W. "The Need for Institutional Overhaul to Permit Conflict Resolution in Energy Policy." Remarks at the World Future Society Energy Forum, Washington, D.C. April 1974.

876. McLaughlin, Donald H. "Man's Selective Attack on Ores and Minerals." In MAN'S ROLE IN CHANGING THE FACE OF THE EARTH, William L. Thomas, Jr., et al. (eds.). The University of Chicago Press, Chicago. 1956. (pp. 851-861).

877. McLean, John G. "U.S. Energy Outlook." Address September 21, 1972. VITAL SPEECHES: November 15, 1972.

878. McLean, John G. "The United States Energy Outlook and Its Implications for National Policy." ANNALS OF THE AMERICAN ACADEMY OF POLITICAL AND SOCIAL SCIENCE 410( ): 97-105. November 1973.

879. McLean, John G. and Davis, Warren B. GUIDE TO NATIONAL PETROLEUM COUNCIL'S REPORT ON U.S. ENERGY OUTLOOK. National Petroleum Council, Washington, D.C. 1973.

880. McMullin, R.J., et al. "Power and the Environment." PUBLIC POWER 28( ): 8-17. 28, 30. May 1970.

881. McNulty, John W. "Utility Aspects of Antipollution Taxes." PUBLIC UTILITIES FORTNIGHTLY 86: 21-24. September 10, 1970.

882. McPhee, John. THE CURVE OF BINDING ENERGY. Farrar, Straus and Giroux, New York. 1974.

883. Macrakis, Michael (ed.). ENERGY: DEMAND, CONSERVATION AND INSTITUTIONAL PROBLEMS. M.I.T. Press, Cambridge, Massachusetts. 1974.

884. McTague, Peter J. "The Environmental War." PUBLIC UTILITIES FORTNIGHTLY 85: 27-35. February 12, 1970.

885. McTague, Peter J. "Energy? Ecology? or Both?" PUBLIC UTILITIES FORTNIGHTLY 87: 15-21. April 1, 1971.

886. McWethy, Patricia J. "Need for Energy Policy Research." POLICY STUDIES JOURNAL 2: 147. 1973.

887. McWethy, Patricia J. "Options for Promotion of Efficient Energy Use." May 1973.

888. Makhijani, A.B. and Lichtenberg, A.J. "Energy and Well-Being." ENVIRONMENT 14( ): 10-18. June 1972.

889. Maler, Karl-Goran. ENVIRONMENTAL ECONOMICS: A THEORETICAL INQUIRY. Published for Resources for the Future by the Johns Hopkins Press, Baltimore. 1974.

890. Malin, H. Martin, Jr. "Toward a National Energy Policy." ENVIRONMENTAL SCIENCE AND TECHNOLOGY 7( ): 392-397. May 1973.

891. Malliaris, A.C. and Strombotne, R.L. "Demand for Energy by the Transportation Sector and Opportunities for Energy Conservation." In ENERGY: DEMAND, CONSERVATION AND INSTITUTIONAL PROBLEMS, Michael Macrakis (ed.). M.I.T. Press, Cambridge, Massachusetts. 1974.

892. Mancke, Richard B. "The Longrun Supply Curve of Crude Oil Produced in the United States." ANTITRUST BULLETIN 15: 727-756. Winter 1970.

893. Mancke, Richard B. "Petroleum Conspiracy: A Costly Myth." PUBLIC POLICY 22(1): 1-14. Winter 1974.

894. Mankoff, Ronald M. and Franklin, William F. "Planning Needed to Overcome Detrimental Effect of New Law on Oil and Gas." JOURNAL OF TAXATION 32: 282-285. May 1970.

895. Mann, Patrick C. "The Impact of Competition in the Supply of Electricity." QUARTERLY REVIEW OF ECONOMICS AND BUSINESS 10: 37-49. Winter 1970.

896. Manners, Gerald. THE GEOGRAPHY OF ENERGY. Aldine Publishing Company, Chicago, Illinois. 1967.

897. Manners, Ian R. and Mikesell, Marvin W. (eds.). PERSPECTIVES ON ENVIRONMENT. Association of American Geographers, Washington, D.C. 1974.

898. Marko, A.M. et al. NUCLEAR POWER AND THE ENVIRONMENT. Atomic Energy of Canada, Ltd., Chalk River, Ontario. 1971.

899. Marts, Marion E. and Sewell, W.R.D. "The Conflict Between Fish and Power Resources in the Pacific Northwest." ANNALS OF THE ASSOCIATION OF AMERICAN GEOGRAPHERS 50(1): 42-50. March 1960. Reprinted in Ian Burton and Robert W. Kates (eds.). READINGS IN RESOURCE MANAGEMENT AND CONSERVATION. The University of Chicago Press, Chicago. 1965. (pp. 327-337).

900. Martz, C.O. "Role of Government in Public Resources Management." ROCKY MOUNTAIN MINERAL AND OIL INSTITUTE 15: 1. 1969.

901. Marx, Wesley. OILSPILL. Sierra Club, San Francisco. 1971.

902. Maryland Academy of Sciences. Study Panel on Nuclear Plants. NUCLEAR POWER PLANTS AND OUR ENVIRONMENT. Report. Maryland Academy of Sciences, Baltimore. 1970.

903. Mason, Edward S. ENERGY REQUIREMENTS AND ECONOMIC GROWTH. National Planning Association, Washington, D.C. 1955.

904. Mason, Peter F. "Some Spatial Implications of a Massive Industrial Accident: The Case of Nuclear Power Plants." THE PROFESSIONAL GEOGRAPHER 24(3): 233-236. August 1972.

905. Mates, Leo. "Oil and the Third World." REVIEW OF INTERNATIONAL AFFAIRS 25(575): 8-9. March 20, 1974.

906. Matson, Roger A. and Studer, Jeannette B. "Energy Resources Development in Wyoming's Powder River Basin: An Assessment of Potential Social and Economic Impacts." Water Resources Research Institute, University of Wyoming, Laramie. April 1974.

907. Matsumura, Seijiro. "'Participation Policy' of the Producing Countries in the International Oil Industry." THE DEVELOPING ECONOMIES 10(1): 30-44. March 1972.

908. Mauthner, Martin U. "The Politics of Energy." EUROPEAN COMMUNITY (174): 13-16. March 1974.

909. Mayer, Lawrence A. "Why the U.S. is in an 'Energy Crisis'." FORTUNE 82: 74-77, 159-160, 162, 164. November 1970.

910. Mazur, Allan. "Cross National Analysis of Electrification: Testing Alternate Theories." Paper Presented at the Annual Meeting of the American Association for the Advancement of Science, San Francisco. February 1974.

911. Mead, Walter J. "The System of Government Subsidies to the Oil Industry." NATURAL RESOURCES JOURNAL 10: 113-125. January 1970.

912. Mead, Walter J. and Sorenson, Philip E. "The Economic Cost of the Santa Barbara Oil Spill." Symposium Proceedings, Marine Science Institute, University of California, Santa Barbara. n.d.

913. Mead, Walter J. and Sorenson, Philip E. "National Defense Petroleum Reserve." LAND ECONOMICS: August 1971.

914. Medvin, Norman. THE ENERGY CARTEL. Random House, New York. 1974. Originally published as THE AMERICAN OIL INDUSTRY, 1973.

915. Meeks, James E. and Landeck, Ronald J. "Area Rate Regulation of the Natural Gas Industry." DUKE LAW JOURNAL 1970: 653-706. August 1970.

916. Meier, Richard L. "Nuclear Agro -- Industrial Complexes for the Future." EKISTICS 26(136): 419-420. November 1968.

917. Meiklejohn, Douglas. "Notes: Liability for Oil Pollution Cleanup and the Water Quality Improvement Act of 1970." CORNELL LAW REVIEW 55( ): 973-991. 1970.

918. Meissner, Tom. "Present and Projected Social Impact Resulting from Coal Development in Seventeen Eastern Montana Counties." Action for Eastern Montana, Glendive, Montana. April 1974.

919. Melamid, Alexander. "Satellization in Iranian Crude-Oil Production." GEOGRAPHICAL REVIEW 63(1); 27-43. January 1973.

920. Melkus, Rolf A. "Toward a Rational Future Energy Policy." NATURAL RESOURCES JOURNAL 14(2): 239-256. April 1974.

921. Merriam, Marshal F. "Decentralized Power Sources for Developing Countries." INTERNATIONAL DEVELOPMENT REVIEW (4): 1972.

922. Metz, W.D. "What Can the Academic Community Do?" SCIENCE 184(4134): 273-275. April 19, 1974.

923. Metzger, H. Peter. THE ATOMIC ESTABLISHMENT. Simon and Schuster, New York. 1972.

924. Michigan Association of School Administrators. Region 9. Energy Conservation Curriculum Committee. ENERGY CONSERVATION: GUIDELINES FOR ACTION. SUGGESTED GUIDELINES FOR LOCAL SCHOOL DISTRICT DEVELOPMENT OF ENERGY CONSERVATION PROGRAMS. Region Nine Superintendents of The Michigan Association of School Administrators, Lansing, Michigan. 1974.

925. MICHIGAN LAW REVIEW. "Jurisdiction -- Atomic Energy -- Federal Pre-Emption and State Regulation of Radioactive Air Pollution: Who is the Master of the Atomic Genie?" MICHIGAN LAW REVIEW 68: 1294-1314. May 1970.

926. MIDDLE EAST INFORMATION SERIES. "The Energy Problem and the Middle East." MIDDLE EAST INFORMATION SERIES 23: 1-96. May 1973.

927. Middle East Institute. THE UNITED STATES AND THE MIDDLE EAST, A RESUME. Middle East Institute, Washington, D.C. 1968.

928. Mikdashi, Zuhayr. A FINANCIAL ANALYSIS OF MIDDLE EASTERN OIL CONCESSIONS: 1901-1965. Praeger, New York. 1966.

929. Mikdashi, Zuhayr. "Cooperation Among Oil Exporting Countries with Special Reference to Arab Countries: A Political Economy Analysis." INTERNATIONAL ORGANIZATION 28(1): 1-30. Winter 1974.

930. Mikesell, Raymond F. and Bartsch, William H. FOREIGN INVESTMENT IN THE PETROLEUM AND MINERAL INDUSTRIES: CASE STUDIES OF INVESTOR-HOST COUNTRY RELATIONS. Published for Resources for the Future by the Johns Hopkins Press, Baltimore. 1971.

-65-

931. Millard, Reed et al. HOW WILL WE MEET THE ENERGY CRISIS? Julian Messner, New York. 1971.

932. Miller, A.J. et al. USE OF STEAM-ELECTRIC POWERPLANTS TO PROVIDE THERMAL ENERGY TO URBAN AREAS. Oak Ridge National Laboratory, Oak Ridge, Tennessee. 1971.

933. Miller, Edward. "Some Implications of Land Ownership Patterns for Petroleum Policy." LAND ECONOMICS 49(4): 414-423. November 1973.

934. Miller, John T., Jr. FOREIGN TRADE IN GAS AND ELECTRICITY IN NORTH AMERICA: A LEGAL AND HISTORICAL STUDY. Praeger, New York. 1970.

935. Miller, John T., Jr. "A Needed Reform of the Organization and Regulation of the Interstate Electric Power Industry." FORDHAM LAW REVIEW 38: 635-673. May 1970.

936. Miller, L.E. "Energy Outlook, Grim Now, But May Ease By 1985." INDUSTRIAL DEVELOPMENT 142( ): 10-12. March-April 1973.

937. Miller, Roger LeRoy. THE ECONOMICS OF ENERGY: WHAT WENT WRONG AND HOW WE CAN FIX IT. Morrow, New York. 1974.

938. Milligan, James. "The North Sea Oil Game." TOWN AND COUNTRY PLANNING 41(7-8): 363-367. July-August 1973.

939. Mills, G. Alex, Johnson, Harry R. and Perry, Harry. "Fuels Management in an Environmental Age." ENVIRONMENTAL SCIENCE AND TECHNOLOGY 5: 30-38. January 1971.

940. Mills, Neil B. "A Look at the Transportation System in the United States." SOCIAL SCIENCE 48(3): 160-166. Summer 1973.

941. Millsap, Ralph H. "Nuclear Energy's Environmental Advantages." EDISON ELECTRIC INSTITUTE BULLETIN 37: 333-336. October 1969.

942. Minick, W. Ted. "Oil and Gas Taxation -- The Dirge of the Abercrombie Doctrine." SOUTHWESTERN LAW JOURNAL 25: 589-595. August 1969.

943. Minks, Merle E. "President Nixon's Energy Message and the Petroleum Industry Lawyer." NATURAL RESOURCES LAWYER 6(4): 513-536. Fall 1973.

944. Mitre Corporation. AN EXAMINATION OF FUEL AND ENERGY INFORMATION SOURCES. Mitre Corporation, McLean, Virginia. 1971.

945. Mitre Corporation. DIMENSIONS OF WORLD ENERGY. The Mitre Corporation, McLean, Virginia. November 1971.

946. Mitre Corporation. ENERGY, RESOURCES AND THE ENVIRONMENT -- MAJOR U.S. POLICY ISSUES. Mitre Corporation, McLean, Virginia. 1972.

947. Mitre Corporation. UNITED STATES TRANSPORTATION: SOME ENERGY AND ENVIRONMENTAL CONSIDERATIONS. Mitre Corporation, McLean, Virginia. 1972.

948. Mitre Corporation. TRANSPORTATION ENERGY AND ENVIRONMENTAL ISSUES. The Mitre Corporation, McLean, Virginia. February 1972.

949. Mitre Corporation. TOWARDS AN ENERGY ETHIC. The Mitre Corporation, McLean, Virginia. March 1972.

950. Mitre Corporation. ENERGY-ENVIRONMENTAL FACTORS IN TRANSPORTATION 1975-1990. The Mitre Corporation, McLean, Virginia. April 1972.

951. Mitre Corporation. SYMPOSIUM ON ENERGY, RESOURCES AND THE ENVIRONMENT. Salines de Chaux Conference Center, Arc-et-Senans, France, May 12-14, 1972. Mitre Corporation, McLean, Virginia. 1972.

952. Mitre Corporation. ECONOMY AND EFFICIENCY IN THE ENERGY SYSTEM. The Mitre Corporation, McLean, Virginia. July 1972.

953. Mitre Corporation. ENVIRONMENTAL ISSUES AND ACTION AROUND THE WORLD. The Mitre Corporation, McLean, Virginia. September 1972.

954. Mitre Corporation. A PRELIMINARY BIBLIOGRAPHY ON ENERGY, RESOURCES AND THE ENVIRONMENT, (KEY-WORD-OUT-OF-CONTEXT LISTING). The Mitre Corporation, McLean, Virginia. October 1972.

955. Mitre Corporation. ENVIRONMENTAL IMPACT OF THE NUCLEAR POWER GENERATING INDUSTRY. The Mitre Corporation, McLean, Virginia. November 1972.

956. Mitre Corporation. ENERGY, RESOURCES AND THE ENVIRONMENT (A SUMMARY OF THE SYMPOSIA). The Mitre Corporation, McLean, Virginia. December 1972.

957. Mitre Corporation. IMPACT OF NEW ENERGY TECHNOLOGY USING GENERALIZED INPUT - OUTPUT ANALYSIS. The Mitre Corporation, McLean, Virginia. February 1973.

958. Mitre Corporation. THE MANAGEMENT OF ENERGY RESEARCH, DEVELOPMENT, AND DEMONSTRATION PROGRAMS: POLICY ALTERNATIVES. The Mitre Corporation, McLean, Virginia. July 1973.

959. Mitre Corporation. THE ECONOMICS OF INTEGRATED ENERGY SYSTEMS. The Mitre Corporation, McLean, Virginia. September 1973.

960. Mitre Corporation. THE FUTURE ROLE OF NUCLEAR ENERGY IN THE UNITED STATES. The Mitre Corporation, McLean, Virginia. September 1973.

-67-

961. Mitre Corporation. ENERGY RELATED ENVIRONMENTAL RESEARCH AGENDA. The Mitre Corporation, McLean, Virginia. December 1973.

962. Mitre Corporation. AN AGENDA FOR RESEARCH AND DEVELOPMENT ON END USE ENERGY CONSERVATION. The Mitre Corporation, McLean, Virginia. (Forthcoming).

963. Mitre Corporation. ENERGY/ENVIRONMENT SCENARIOS 1975-2000. The Mitre Corporation, McLean, Virginia. (Forthcoming).

964. Mohring, Herbert. "Optimization and Scale Economies in Urban Bus Transportation." THE AMERICAN ECONOMIC REVIEW 62(4): 591-604. September 1972.

965. Molotch, Harvey. "Oil in the Velvet Playground." RAMPARTS 8: 44-51. 1970.

966. Molotch, Harvey. "Oil in Santa Barbara and Power in America." SOCIOLOGICAL INQUIRY 40(1): 131-144. Winter 1970.

967. Montana, University of. Institute for Social Science Research. "A Comparative Case Study of the Impact of Coal Development on the Way of Life of People in the Coal Areas of Eastern Montana and Northeastern Wyoming." Institute for Social Science Research, University of Montana, Missoula. April 1974.

968. Montgomery, Suzanne. "Oil Pollution R&D on the Rise." UNDERSEA TECHNOLOGY 10: 30-31, 48-51. October 1969.

969. Mooney, R.E. "Energy and Lawlessness." NEW YORK TIMES: Feburary 17, 1974.

970. Moore, David D., Gaines, Gordon B. and Hessel, Darryl. "Getting Energy to the User." BATTELLE RESEARCH OUTLOOK 4(1): 1972.

971. Moore, Thomas Gale. "The Effectiveness of Regulation of Electric Utility Prices." SOUTHERN ECONOMIC JOURNAL 36( ): 365-375. April 1970.

972. Mooz, W.E. AN EXAMINATION OF A NATIONAL ENERGY STUDY. National Science Foundation, Washington, D.C. 1970.

973. Mooz, W.E. SOME FACTS ON THE SUPPLY OF ENERGY. National Science Foundation, Washington, D.C. 1970.

974. Morgan, David P. "Oil by Rail." ENVIRONMENT 14(8): 30-31. October 1972.

975. Morgan, Thomas. ATOMIC ENERGY AND CONGRESS. University of Michigan Press, Ann Arbor. 1956.

976. Morrell, Gene P. "Federal Regulation of Energy Production." PROCEEDINGS OF THE ACADEMY OF POLITICAL SCIENCE 31(2): 159-169. December 1973.

977. Morris, Deane M. FUTURE ENERGY DEMAND AND ITS EFFECT ON THE ENVIRONMENT. Rand Corporation, Santa Monica, California. 1972.

978. Morris, Peter A. "Power Plant Reactor Safety and Risk Approval." In ENERGY, THE ENVIRONMENT AND HUMAN HEALTH, Asher J. Finkel (ed.). Publishing Sciences Group, Acton, Massachusetts. 1973.

979. Morrison, David L., Erb, Donald E. and Reid, William T. ENERGY IN THE URBAN ENVIRONMENT. Paper Number 71-526. American Institute of Aeronautics and Astronautics. 1971.

980. Morrison, Denton E. "Declining Resources, Discontent and Some Imperatives for the Future." Paper Presented at the Annual Meeting of the Rural Sociological Society, Montreal, Canada. August 1974.

981. Morrison, Denton E., Hornback, Kenneth E. and Warner, W. Keith. ENVIRONMENT: A BIBLIOGRAPHY OF SOCIAL SCIENCE AND RELATED LITERATURE. U.S. Environmental Protection Agency. U.S. Government Printing Office, Washington, D.C. 1974.

982. Morrison, Warren E. AN ENERGY MODEL FOR THE UNITED STATES. U.S. Government Printing Office, Washington, D.C. 1968.

983. Morton, Rogers C.B. "Strip-Mining Reform -- Some Political and Economic Ideas." ENVIRONMENTAL AFFAIRS 2(2): 294-302. Fall 1972.

984. Morton, Rogers C.B. "The Nixon Adminstration Energy Policy." ANNALS OF THE AMERICAN ACADEMY OF POLITICAL AND SOCIAL SCIENCE 410( ): 65-74. November 1973.

985. Mosley, Leonard. POWER PLAY: OIL IN THE MIDDLE EAST. Random House, New York. 1973.

986. Moxness, R. "The Long Pipe." ENVIRONMENT 12( ): 12-23. September 1970.

987. Muir, J. Dapray. "Licensing of Nuclear Power Reaction in the United States -- New Developments." ATOMIC ENERGY LAW JOURNAL 15( ): 135- . 1973.

988. Muir, J. Dapray. "The Environmentalist's View of AEC's Judicial Function. ATOMIC ENERGY LAW JOURNAL 15( ): 176- . 1973.

989. Muir, J. Dapray. "Legal and Ecological Aspects of the International Energy Situation." THE INTERNATIONAL LAWYER 8( ): 1-10. January 1974.

-69-

990. Mullan, Joseph W. "Fossil Fuels." In ENERGY, THE ENVIRONMENT, AND HUMAN HEALTH, Asher J. Finkel (ed.). Publishing Sciences Group, Acton, Massachusetts. 1973.

991. Mullenbach, Philip. CIVILIAN NUCLEAR POWER: ECONOMIC ISSUES AND POLICY FORMATION. Twentieth Century Fund, New York. 1963.

992. Muntzing, L. Manning. Director of Regulation, U.S. Atomic Energy Commission. Testimony on "Siting and Licensing of Nuclear Power Plants" before the Joint Committee on Atomic Energy. March 1974.

993. Murdoch, William W. ENVIRONMENT: RESOURCES, POLLUTION AND SOCIETY. Sinauer Associates, Stamford, Connecticut. 1971.

994. Murphy, Earl Finbar. "The Effect of Law, Economics and Politics on Energy Resources Development." CASE WESTERN RESERVE JOURNAL OF INTERNATIONAL LAW 5(1): 81-86. Winter 1972.

995. Murphy, John J. (ed.). ENERGY AND PUBLIC POLICY -- 1972. Report Number 575. The Conference Board, New York. 1972.

996. Murray, J.R. "Energy-Related Data: Results for the Period May 3 - May 30, 1974." Continuous National Survey, National Opinion Research Center, Chicago, Illinois. June 1974.

997. Murray, J.R. et al. "Evolution of Public Response to the Energy Crisis." SCIENCE 184(4134): 257-263. April 19, 1974.

998. Musgrove, Peter J. and Wilson, Alan D. "Power Without Pollution." NEW SCIENTIST 45: 457-459. March 5, 1970.

999. Muskie, Edmund S. "Power and Pollution." RURAL ELECTRIFICATION 9( ): 27-28, 46. June 1970.

1000. Myers, John G. "Energy: The Next Six Months." THE CONFERENCE BOARD RECORD 11(3): 10-13. March 1974.

1001. Myers, John G. "Energy Use by Industry." THE CONFERENCE BOARD RECORD 11(5): 32-36. May 1974.

1002. Nader, Ralph. "Nuclear Power on Trial." TRIAL 10(1): 19-25. January-February 1974.

1003. Nanda, Ved P. "The TORREY CANYON Disaster: Some Legal Aspects." DENVER LAW JOURNAL 44( ): 400-425. 1967.

1004. Nassikas, John N. "Regulation of the Electric Utilities in the 1970's." EDISON ELECTRIC INSTITUTE BULLETIN: 214-218, 254. July-August 1970.

1005. Nassikas, John N. "Coordination of Electrical Power and Environmental Policy." NATURAL RESOURCES LAWYER 4( ): 268- . 1971.

1006. Nassikas, John N. "National Energy and Environmental Policy." PUBLIC UTILITIES FORTNIGHTLY 87( ): 49-61. June 10, 1971.

1007. Nassikas, John N. "Energy Policy Imperatives." NATURAL RESOURCES LAWYER 5: 627. Fall 1972.

1008. National Academy of Engineering. ENGINEERING FOR RESOLUTION OF THE ENERGY-ENVIRONMENT DILEMMA. National Academy of Engineering, Washington, D.C. 1971.

1009. National Academy of Engineering. U.S. ENERGY PROSPECTS. National Academy of Engineering, Washington, D.C. 1974.

1010. National Academy of Sciences. RESOURCES AND MAN: A STUDY AND RECOMMENDATIONS. National Academy of Sciences, Washington, D.C. 1969.

1011. National Academy of Sciences. SOLAR ENERGY IN DEVELOPING COUNTRIES: PERSPECTIVES AND PROSPECTS. National Academy of Sciences, Washington, D.C. 1972.

1012. National Academy of Sciences. ENVIRONMENTAL QUALITY AND SOCIAL BEHAVIOR: STRATEGIES FOR RESEARCH. National Academy of Sciences, Washington, D.C. 1973.

1013. National Academy of Sciences. REHABILITATION POTENTIAL OF WESTERN COAL LANDS. National Academy of Sciences, Washington, D.C. 1974.

1014. National Center for Resource Recovery. "Municipal Solid Waste -- A Source of Energy." NCRR BULLETIN: Summer 1973.

1015. National Coal Association. WHY IS COAL IN VERY TIGHT SUPPLY? WHAT CAN BE DONE ABOUT IT? National Coal Association, Washington, D.C. 1971.

1016. National Economic Research Associates, Inc. ENERGY CONSUMPTION AND GROSS NATIONAL PRODUCT IN THE UNITED STATES: AN EXAMINATION OF A RECENT CHANGE IN THE RELATIONSHIP. National Economic Research Associates, Inc., Washington, D.C. 1971.

1017. National Electrical Manufacturers Association. SECOND BIENNIAL SURVEY OF POWER EQUIPMENT REQUIREMENTS OF THE U.S. ELECTRIC UTILITY INDUSTRY -- 1969-78. National Electrical Manufacturers Association, New York. 1970.

1018. National Energy Board. ENERGY SUPPLY AND DEMAND IN CANADA AND EXPORT DEMAND FOR CANADIAN ENERGY 1960-1990. The Queen's Printer, Ottawa. 1969.

1019. National Energy Forum. ENERGY POLICY ISSUES FOR THE UNITED STATES DURING THE SEVENTIES. Prepared by Arthur D. Little, Inc., Arlington, Virginia. 1971.

1020. NATIONAL JOURNAL REPORTS. "Energy Policies and Problems." NATIONAL JOURNAL REPORTS: October 13, 1973.

1021. National Materials Advisory Board. ELEMENTS OF A NATIONAL MATERIALS POLICY. National Materials Advisory Board, Washington, D.C. 1972.

1022. National Petroleum Council. FACTORS AFFECTING U.S. EXPLORATION, DEVELOPMENT AND PRODUCTION, 1946-1965. National Petroleum Council, Washington, D.C. 1967.

1023. National Petroleum Council. U.S. PETROLEUM AND GAS TRANSPORTATION CAPACITIES. National Petroleum Council, Washington, D.C. 1967.

1024. National Petroleum Council. PETROLEUM RESOURCES UNDER THE OCEAN FLOOR. National Petroleum Council, Washington, D.C. 1969.

1025. National Petroleum Council. U.S. PETROLEUM IMPORTS. National Petroleum Council, Washington, D.C. 1969.

1026. National Petroleum Council. FUTURE PETROLEUM PROVINCES OF THE UNITED STATES: A SUMMARY. National Petroleum Council, Washington, D.C. 1970.

1027. National Petroleum Council. U.S. PETROLEUM INVENTORIES AND STORAGE CAPACITY. National Petroleum Council, Washington, D.C. 1970.

1028. National Petroleum Council. ENVIRONMENTAL CONSERVATION: THE OIL AND GAS INDUSTRIES. National Petroleum Council, Washington, D.C. 1971.

1029. National Petroleum Council. OIL AND GAS TRANSPORTATION FACILITIES. National Petroleum Council, Washington, D.C. 1972.

1030. National Petroleum Council. U.S. ENERGY OUTLOOK. National Petroleum Council, Washington, D.C. 1972.

1031. National Petroleum Council. Committee on Crude Oil Deliverability. CAPACITY OF CRUDE OIL GATHERING SYSTEMS AND DEEPWATER TERMINALS. National Petroleum Council, Washington, D.C. 1970.

1032. National Petroleum Council. Committee on Emergency Fuel Convertibility. EMERGENCY FUEL CONVERTIBILITY. National Petroleum Council, Washington, D.C. 1965.

1033. National Petroleum Council. Committee on Fuels. SHORT-TERM FUEL OUTLOOK. National Petroleum Council, Washington, D.C. 1970.

1034. National Petroleum Council. Committee on Future Petroleum and Gas Producing Capabilities. ESTIMATED PRODUCTIVE CAPACITIES OF CRUDE OIL, NATURAL GAS AND NATURAL GAS LIQUIDS IN THE UNITED STATES (1965-1970). National Petroleum Council, Washington, D.C. 1966.

1035. National Petroleum Council. Committee on National Oil Policy. PETROLEUM POLICIES FOR THE UNITED STATES. National Petroleum Council, Washington, D.C. 1966.

1036. National Petroleum Council. Committee on U.S. Energy Outlook. U.S. ENERGY OUTLOOK: AN INITIAL APPRAISAL, 1971-1985; AN INTERIM REPORT. National Petroleum Council, Washington, D.C. 1971.

1037. National Power Survey. Technical Advisory Committee on Transmission. THE TRANSMISSION OF ELECTRIC POWER. U.S. Federal Power Commission, Washington, D.C. 1970.

1038. National Research Council. Task Group T-65 on Air Pollution. IMPACT OF AIR POLLUTION REGULATIONS ON FUEL SELECTION FOR FEDERAL FACILITIES. Technical Report Number 57. National Academy of Sciences, Washington, D.C. 1970.

1039. National Rural Electric Cooperative Association, Research Division. DIMENSIONS OF THE NATIONAL POWER CRISIS: BASIC INFORMATION IN QUESTION - AND - ANSWER FORM FOR THE CONSUMER-OWNERS, OFFICERS, DIRECTORS AND STAFFS OF RURAL ELECTRIC UTILITY SYSTEMS, Research Paper Number 71-3. National Rural Electric Cooperative Association, Washington, D.C. 1971.

1040. National Science Foundation. SUMMARY REPORT OF THE CORNELL WORKSHOP ON ENERGY AND THE ENVIRONMENT. U.S. Government Printing Office, Washington, D.C. 1972.

1041. National Science Foundation and NASA Solar Energy Panel. ASSESSMENT OF SOLAR ENERGY AS A NATIONAL ENERGY RESOURCE. U.S. Government Printing Office, Washington, D.C. 1972.

1042. NATIONAL WILDLIFE. "Nuclear Power Plants: Boon or Blight?" NATIONAL WILDLIFE: April/May 1971.

1043. National Wildlife Federation. "National Academy of Sciences Report on Western Coal Stripping and Reclamation." Report Number 31. CONSERVATION REPORT. September 1973.

1044. NATURAL RESOURCES LAWYER. "Petroleum Conservation -- How America is Making the Most of its Oil and Gas Resources." NATURAL RESOURCES LAWYER 3: 272. May 1970.

1045. NATURE (eds.). "Energy Review." NATURE 249(5459): 697-737. June 21, 197

1046. NATURE (eds.). "Rising Tide of Energy Nationalism." NATURE 249(5459): 681. June 21, 1974.

1047. Nef, John U. THE RISE OF THE BRITISH COAL INDUSTRY. Routledge, London. 1932.

1048. Nef, John U. THE CONQUEST OF THE MATERIAL WORLD. University of Chicago Press, Chicago. 1964.

1049. Negre, D.V. and Adler, J.H. "The Ecology of Terrorism." SURVIVAL: 178, 181-183. July/August 1973.

1050. Nelkin, Dorothy. NUCLEAR POWER AND ITS CRITICS: THE CAYUGA LAKE CONTROVERSY. Cornell University Press, Ithaca, New York. 1971.

1051. Nelkin, Dorothy. "Experts and Citizens: Problems of Participation in Power Plant Siting Decisions." Paper Presented at the Annual Meeting of the American Association for the Advancement of Science, San Francisco. February 1974.

1052. Nelkin, Dorothy. "Technical Expertise as a Political Resource: A Power Plant Siting Controversy." BULLETIN OF THE ATOMIC SCIENTISTS. September 1974. (Forthcoming).

1053. Nelkin, Dorothy. "The Political Impact of Technical Expertise." SOCIAL STUDIES OF SCIENCE. January 1975. (Forthcoming).

1054. Nelkin, Dorothy and Kupchak, Kenneth P. "The Legal Setting of Nuclear Powerplant Siting Decisions: A New York State Controversy." CORNELL LAW REVIEW 57(1): 80-104. November 1971.

1055. Nellis, Lee. "What Does Energy Development Mean for Wyoming? A Community Study at Hanna, Wyoming." Office of Special Projects. University of Wyoming, Laramie. n.d.

1056. Nelson, Jon and Ferrar, Terry A. "An Economic Analysis of Recent Government Energy Policy: Residential and Commercial Heating Oil Demand Modifications." Pennsylvania State University, University Park. 1974.

1057. Nelson, L.F. and Burrows, W.C. "Putting the U.S. Agricultural Energy Picture into Focus." Paper Presented at the Annual Meeting of the American Society of Agriculture Engineers. June 1974. ASAE Paper Number 74-1040.

1058. Netschert, Bruce C. "Competition in the Nuclear Power Supply Industry." With reply by Jesse W. Markham. ANTITRUST BULLETIN 14: 629-666. Fall 1969.

1059. Netschert, Bruce C. "The Economic Impact of Electric Vehicles: A Scenario." BULLETIN OF THE ATOMIC SCIENTISTS 26: 29-35. May 1970.

1060. Netschert, Bruce C. "The Energy Company: A Monopoly Trend in the Energy Markets." BULLETIN OF THE ATOMIC SCIENTISTS 27: October 1971

1061. Netschert, Bruce C. "Avoiding Platitudes and Panaceas Is Also a Policy Problem." THE CONFERENCE BOARD RECORD 9(7): 25-26. July 1972.

1062. Netschert, Bruce C. "Are Antipollution Laws Widening the Energy Gap?" In ENERGY AND THE ENVIRONMENT: A COLLISION OF CRISES, Irwin Goodwin (ed.). Publishing Sciences Group, Acton, Massachusetts. 1973.

1063. Netschert, Bruce C. "Energy Versus Environment." HARVARD BUSINESS REVIEW 51(1): 24-28, 133-140. January-February 1973.

1064. Netschert, Bruce C., Gerber, Abraham and Stelzer, I.M. COMPETITION IN THE ENERGY MARKETS, AN ECONOMIC ANALYSIS. National Economic Research Associates, Washington, D.C. 1970.

1065. Neuner, Edward J. THE NATURAL GAS INDUSTRY. University of Oklahoma Press, Tulsa. 1960.

1066. Nevins, Allan et al. ENERGY AND MAN: A SYMPOSIUM. Appleton-Century-Crofts, New York. 1960.

1067. Newman, Dorothy K. and Wachtel, Dawn Day. "Energy, the Environment and the Poor." Paper Presented at the Annual Meeting of the Society for the Study of Social Problems, Montreal. August 1974.

1068. NEW SCIENTIST. "Energy for the World's Technology." NEW SCIENTIST 44: 1-24. November 13, 1969.

1069. New York Bar Association. Special Committee on Electric Power and the Environment. ELECTRICITY AND THE ENVIRONMENT: THE REFORM OF LEGAL INSTITUTIONS. West Publishing Co., St. Paul, Minnesota. 197

1070. New York City Environmental Protection Administration. TOWARD A RATIONAL POWER POLICY: RECONCILING NEEDS FOR ENERGY AND ENVIRONMENTAL PROTECTION. Report to Mayor's Interdepartmental Committee on Public Utilities. New York City Environmental Protection Admininistration New York. 1971.

1071. New York (State) Department of Public Service. Gas Division. GAS SUPPLIES FOR U.S. CONSUMERS. New York State Department of Public Service, Albany, New York. 1971.

1072. Nichols, Barry L. "Waste-Heat Utilization." PROCEEDINGS OF THE ACADEMY OF POLITICAL SCIENCE 31(2): 87-97. December 1974.

1073. Nicholson, R.L.R. "The Nuclear Power Paradox in the UK." ENERGY POLICY 1(1): 38-46. June 1973.

1074. Nicholson, R.L.R. "Plutonium and the Fast Reactor." ENERGY POLICY 2(2): 159. June 1974.

1075. Noble, David. "Interlocking Fuel Interests." ENERGY DIGEST 1: 27-37. April 13, 1971.

1076. Noggle, Burl. TEAPOT DOME: OIL AND POLITICS IN THE 1920's. Louisiana State University Press, Baton Rouge. 1962.

1077. Noll. "The Economics and Politics of Regulation." VIRGINIA LAW REVIEW 57( ): 1016- . 1971.

1078. Noone, James A. "Oil Import Needs vs. Environmental Costs: Key Issue in Deep Water Ports Legislation." NATIONAL JOURNAL REPORTS: November 10, 1973.

1079. Noone, James A. "Environment and Energy." NATIONAL JOURNAL REPORTS 5(51): 1911-1930. December 22, 1973.

1080. Nordhaus, William D. "The Allocation of Energy Resources." BROOKINGS PAPERS ON ECONOMIC ACTIVITY (3): 529-576. 1973.

1081. Nordhauser, Norman. "Origins of Federal Oil Regulation in the 1920's." BUSINESS HISTORY REVIEW 47(1): 53-71. Spring 1973.

1082. North, D.C. and Miller, R.L. THE ECONOMICS OF PUBLIC ISSUES. Harper and Row, New York. 1971.

1083. Northern Great Plains Resources Program. Work Group F. "Socio-Economic and Cultural Aspects of Potential Coal Development in the Northern Great Plains." Discussion Draft. Northern Great Plains Resources Program, Denver, Colorado. June 1974.

1084. Novick, Sheldon. THE CARELESS ATOM. Houghton Mifflin, Boston. 1969.

1085. Novick, Sheldon. "Toward a Nuclear Power Precipice." ENVIRONMENT 15(2): 32-40. March 1973.

1086. Novick, Sheldon. "Looking Forward: Saving the Environment Through More Effective Use of Energy." ENVIRONMENT 15: 4-15. May 1973.

1087. Novick, Sheldon. "Censoring Nuclear Debate." ENVIRONMENT 16(4): 14-17. April 1974.

1088. Novick, Sheldon. "Nuclear Breeders." ENVIRONMENT 16(6): 6-15. July-August 1974.

1089. Oak Ridge National Laboratory. NUCLEAR ENERGY CENTERS: INDUSTRIAL AND AGRO-INDUSTRIAL COMPLEXES, SUMMARY REPORT. Oak Ridge National Laboratory, Oak Ridge, Tennessee. 1970.

1090. Oak Ridge National Laboratory. CIVILIAN NUCLEAR POWER: REACTOR FUEL CYCLE COSTS FOR NUCLEAR POWER EVALUATION. Oak Ridge National Laboratory, Oak Ridge, Tennessee. 1971.

1091. Oak Ridge National Laboratory. SURVEY OF THERMAL RESEARCH PROGRAMS SPONSORED BY FEDERAL, STATE AND PRIVATE AGENCIES. Oak Ridge National Laboratory, Oak Ridge, Tennessee. 1971.

1092. Oak Ridge National Laboratory. ENERGY CONSUMPTION FOR TRANSPORTATION IN THE U.S. National Technical Information Service, Springfield, Virginia. 1972.

1093. Oak Ridge National Laboratory. AN INVENTORY OF ENERGY RESEARCH, PREPARED FOR THE TASK FORCE ON ENERGY. U.S. Government Printing Office, Washington, D.C. 1972.

1094. Oak Ridge National Laboratory. USE OF STEAM-ELECTRIC POWER PLANTS TO PROVIDE THERMAL ENERGY TO URBAN AREAS. National Technical Information Service, Springfield, Virginia. 1972.

1095. O'Connor, Harvey. WORLD CRISIS IN OIL. Monthly Review Press, New York. 1962.

1096. O'Connor, Lawrence J., Jr. "Gas Supply and the Role of the Independent Producer." PUBLIC UTILITIES FORTNIGHTLY 84: 26-33. October 23, 1969.

1097. Odell, Peter R. OIL AND WORLD POWER: A GEOGRAPHICAL INTERPRETATION. Cox and Wyman, Ltd., London. 1970.

1098. Odell, Peter R. "Indigenous Oil and Gas Developments and Western Europe's Energy Policy Options." ENERGY POLICY 1(1): 47-64. June 1973.

1099. Odell, Peter R. "Energy -- From Surplus to Scarcity?" ENERGY POLICY 1(2): 164-165. September 1973.

1100. Odell, Peter R. "The Future of Oil: A Rejoinder." THE GEOGRAPHICAL JOURNAL 139(3): 436-454. October 1973.

1101. Odell, Peter R. "Europe Sits on its Own Energy: Oil for the 1980's." GEOGRAPHICAL MAGAZINE 46(6): 241-245. March 1974.

1102. Odell, Peter R. "World Energy in the Balance: Self-Sufficiency for Economic Survival." GEOGRAPHICAL MAGAZINE 46(8): 378-379. May 1974.

1103. Odum, Howard T. "Energetics of World Food Production." In THE WORLD FOOD PROBLEM, A REPORT OF THE PRESIDENT'S SCIENCE ADVISORY COMMITTEE, Volume 3. U.S. Government Printing Office, Washington, D.C. 1967. (pp. 55-94).

1104. Odum, Howard T. ENVIRONMENT, POWER AND SOCIETY. Wiley-Interscience, New York. 1971.

1105. Odum, Howard T. "Terminating Fallacies in National Policy on Energy, Economics, and Environment." In ENERGY: TODAY'S CHOICES, TOMORROW'S OPPORTUNITIES, Anton B. Schmalz (ed.). World Future Society, Washington, D.C. 1974. (pp. 15-19).

1106. Odum, Howard T. "Energy, Ecology and Economics." THE MOTHER EARTH NEWS: 6-12. May 1974.

1107. Odum, Howard T. "More Perspectives on World Energy Relationships." THE MOTHER EARTH NEWS: 12-13. May 1974.

1108. Ogle, Richard A. "Institutional Factors to Encourage Interagency Cooperation in the Management of Natural Resources." PUBLIC ADMINISTRATION REVIEW: January/February 1972.

1109. Olniensis, S. et al. ELECTRICITY AND THE ENVIRONMENT: THE REFORM OF LEGAL INSTITUTIONS. West Publishing Company, Saint Paul, Minnesota. 1972.

1110. Olson, Charles E. COST CONSIDERATIONS FOR EFFICIENT ELECTRICITY SUPPLY. Michigan State University, East Lansing. 1970.

1111. Olson, Charles E. and Cumberland, John H. "Bulk Power and Environmental Pollution." PUBLIC UTILITIES FORTNIGHTLY 85: 21-27. May 7, 1970.

1112. Organization for Economic Cooperation and Development. "Some Aspects of United States Energy Policy." OECD OBSERVER 28: 27-31. June 1967.

1113. Organization for Economic Cooperation and Development. ENERGY POLICY: PROBLEMS AND OBJECTIVES. Organization for Economic Co-Operation and Development, Paris. 1968.

1114. Organization for Economic Cooperation and Development. ORGANIZATION AND GENERAL REGIME GOVERNING NUCLEAR ACTIVITIES. European Nuclear Energy Agency, Paris. 1969.

1115. Organization for Economic Cooperation and Development. REPORTS BY MEMBER COUNTRIES ON THE ADMINISTRATIVE STRUCTURE FOR DEALING WITH ENERGY PROBLEMS. Organization for Economic Co-Operation and Development, Paris. 1969.

1116. Organization for Economic Cooperation and Development. "Sweden's Energy Policy." OECD OBSERVER (40): 43-45. June 1969.

1117. Organization for Economic Cooperation and Development. "Natural Gas: Its Impact on the Energy Market in OECD Europe." OECD OBSERVER (42): 37-41. October 1969.

1118. Organization for Economic Cooperation and Development. "The Energy Policy of Japan." OECD OBSERVER 48: 15-18. October 1970.

1119. Organization for Economic Cooperation and Development. OIL, THE PRESENT SITUATION AND FUTURE PROSPECTS. Organization for Economic Cooperation and Development, Paris. 1973.

1120. Organization for Economic Cooperation and Development. "Changing Role for OECD's Nuclear Energy Agency." ORGANIZATION FOR ECONOMIC COOPERATION AND DEVELOPMENT OBSERVER (66): 19-26. October 1973.

1121. Organization for Economic Cooperation and Development. "OECD and the Energy Problem." OECE OBSERVER (66): 7-8. October 1973.

1122. Organization for Economic Cooperation and Development. Energy Committee. ENERGY POLICY: PROBLEMS AND OBJECTIVES. Organization for Economic Cooperation and Development, Paris. 1966.

1123. Osborn, Elburt F. "Coal and the Present Energy Situation." SCIENCE 183( ): 477-481. February 8, 1974.

1124. Osterhoudt, Frank H. "Social and Economic Impacts From Strip Coal Mining in the Great Plains." Paper Presented at the Meetings of the Great Plains Agricultural Council, Lincoln, Nebraska. 1974.

1125. Osterhoudt, Frank H. "Social and Economic Impact of Potential Coal Development in the Northern Great Plains." Natural Resource Economics Division, Economic Research Service, U.S. Department of Agriculture, Washington, D.C. 1974.

1126. Overseas Development Council. THE UNITED STATES AND THE DEVELOPING WORLD: AGENDA FOR ACTION, 1973. Overseas Development Council, Washington, D.C. 1973.

1127. Owen, Bruce M. "Monopoly Pricing in Combined Gas and Electric Utilities. ANTITRUST BULLETIN 15: 713-726. Winter 1970.

1128. Ouellette, Robert P. "Energy and Environmental Quality." PROCEEDINGS OF THE ACADEMY OF POLITICAL SCIENCE 31(2): 170-182. December 1973.

1129. Ouellette, Robert P. BIBLIOGRAPHY OF RECENT MITRE WORK IN ENERGY, RESOURCES AND THE ENVIRONMENT. The Mitre Corporation, McLean, Virginia. 1974.

1130. Owen, Oliver S. NATURAL RESOURCE CONSERVATION. Macmillan, New York. 1971.

1131. Pace, Clark. "When Built-In Growth Strikes Back." EXCHANGE 30: 6-13. October 1969.

1132. Pace, Joe D. "The Subsidy Received by Publicly Owned Electric Utilities." PUBLIC UTILITIES FORTNIGHTLY: April 29, 1971.

1133. Packer, Arnold H. "The Noneconomics of Energy Policy." Paper Presented at the Annual Meeting of the American Association for the Advancement of Science, San Francisco. February 1974.

1134. Page, Donald M. "Energy Crisis in Reverse." INTERNATIONAL PERSPECTIVES: 18-21. March-April 1974.

1135. Park, Charles Frederick, Jr. AFFLUENCE IN JEOPARDY: MINERALS AND THE POLITICAL ECONOMY. W.H. Freeman, Cooperation. San Francisco. 1968.

1136. Passer, Harold. THE ELECTRICAL MANUFACTURERS 1875-1900. Harvard University Press, Cambridge, Massachusetts. 1953.

1137. Passmore, John. MAN'S RESPONSIBILITY FOR NATURE: ECOLOGICAL PROBLEMS AND WESTERN TRADITIONS. Charles Scribner's Sons, New York. 1974.

1138. Paulsen, David F. and Denhardt, Robert B. (eds.). POLLUTION AND PUBLIC POLICY. Dodd, Mead, New York. 1973.

1139. Pavelis, George A. ENERGY, NATURAL RESOURCES AND RESEARCH IN AGRICULTURE: EFFECTS ON ECONOMIC GROWTH AND PRODUCTIVITY FOR THE UNITED STATES. Economic Research Service, U.S. Department of Agriculture, Washington, D.C. August 1973.

1140. Peach, W.N. ENERGY OUTLOOK FOR THE 1980's. Study Prepared for the Subcommittee on Economic Progress of the Economic Joint Committee. U.S. Congress. U.S. Government Printing Office, Washington, D.C. 1973.

1141. Pecora, William T. "The Administration's Energy Message and Program." THE CONFERENCE BOARD RECORD 9(7): 27-30. July 1972.

1142. Pederson, William F., Jr. "Are Antipollution Laws Working?" In ENERGY AND THE ENVIRONMENT: A COLLISION OF CRISES, Irwin Goodwin (ed.). Publishing Sciences Group, Acton, Massachusetts. 1973.

1143. Peele, Elizabeth. "Social Effects of Nuclear Power Plants." Oak Ridge National Laboratory, Oak Ridge, Tennessee. n.d.

1144. Pennsylvania Department of Education. ENVIRONMENTAL IMPACT OF ELECTRICAL POWER GENERATION, NUCLEAR AND FOSSIL. MINICOURSE FOR SECONDARY SCHOOLS AND ADULT EDUCATION. Prepared for the Atomic Energy Commission. U.S. Government Printing Office, Washington, D.C. 1973.

1145. Penrose, Edith. "The Oil 'Crisis'." THE ROUND TABLE (254): 135-148. April 1974.

1146. Penzin, D. "New Moves by Oil Imperialism." INTERNATIONAL AFFAIRS (MOSCOW) 5: 47-53. May 1973.

1147. Perelman, Michael J. "Farming with Petroleum." ENVIRONMENT 14(8): 8-13. October 1972.

1148. Perelman, Michael. "A Minority Report on the Economics of Spatial Heterogeneity in Agricultural Enterprises." In MONOCULTURE IN AGRICULTURE: EXTENT, CAUSES, AND PROBLEMS, U.S. Department of Agriculture, Washington, D.C. 1973. (pp. 53-64).

1149. Perelman, Michael J. and Shea, Kevin P. "The Big Farm." ENVIRONMENT 14( ): 10-15. December 1972.

1150. Permar, David H. "A Legal Solution to the Electric Power Crisis: Controlling Demand Through Regulation of Advertising, Promotion and Rate Structure." ENVIRONMENTAL AFFAIRS 1(3).

1151. Perry, Harry. A REVIEW OF ENERGY ISSUES AND THE 91st CONGRESS. U.S. Government Printing Office, Washington, D.C. 1971.

1152. Perry, Harry. CONSERVATION OF ENERGY. U.S. Government Printing Office, Washington, D.C. 1972.

1153. Perry, Harry. "Goals for an Energy Policy." THE CONFERENCE BOARD RECORD 9(7): 19. July 1972.

1154. Perry, Harry. "How Much Energy Do We Waste?" In ENERGY AND THE ENVIRONMENT: A COLLISION OF CRISES, Irwin Goodwin (ed.). Publishing Sciences Group, Acton, Massachusetts. 1973.

1155. Perry, Harry and Berkson, Harold. "Must Fossil Fuels Pollute?" TECHNOLOGY REVIEW 74: 34-43. December 1971.

1156. Perry, Harry and Weidenfeld, Edward L. SELECTED READINGS ON THE FUELS AND ENERGY CRISIS. U.S. Government Printing Office, Washington, D.C. 1972.

1157. Peterson, Bill. COALTOWN REVISITED: AN APPALACHIAN NOTEBOOK. Henry Regnery Company, Chicago. 1972.

1158. Peterson, J.A. "Rape of a Wilderness Lake: Hydro-Electric Power Invades Tasmania's Landscape Heritage." THE GEOGRAPHICAL MAGAZINE 45(5): 371-376. February 1973.

1159. Peterson, Russell W. "Consumption with Conservation -- Can We Have Both?" THE CONFERENCE BOARD RECORD 11(5): 29-31. May 1974.

1160. Petrie, D. PETROLEUM. Oxford University Press, London. 1961.

1161. Petroleum Industry Research Foundation. THE TAX BURDEN ON THE DOMESTIC OIL AND GAS INDUSTRY, 1962-1963. Petroleum Industry Research Foundation, New York. 1965.

1162. Petroleum Industry Research Foundation. ECONOMIC/SOCIAL IMPLICATIONS OF REMOVING THE PERCENTAGE DEPLETION PROVISION. Petroleum Industry Research Foundation, New York. 1969.

1163. Petroleum Industry Research Foundation. OIL IMPORTS AND THE NATIONAL INTEREST. Petroleum Industry Research Foundation, New York. 1971.

1164. Petroleum Publishing Company. INTERNATIONAL PETROLEUM ENCYCLOPEDIA -- 1970. Petroleum Publishing Company, Tulsa, Oklahoma. 1969.

1165. Petrow, Richard. IN THE WAKE OF TORREY CANYON. McKay, New York. 1968.

1166. Pfeiffer, Brant A. and Gilbert, Ronald D. "Pollution Abatement Expenditures by the Electric Power Industry." PUBLIC UTILITIES FORTNIGHTLY 90(5): 21-28. August 31, 1972.

1167. Phillips, H.H. "President Nixon's Energy Message and the Electric Generating Industry Lawyer." NATURAL RESOURCES LAWYER 6(4): 537-542. Fall 1973.

1168. Phillips, James G. "Fuel Shortages." EDITORIAL RESEARCH REPORTS 2(14): 755-772. 1970.

1169. Phillips, James G. "Natural Gas Deregulation." NATIONAL JOURNAL REPORTS 6(21): 761-775. May 25, 1974.

1170. Phillips, James G. "U.S. Role in Foreign Oil Talks." NATIONAL JOURNAL REPORTS 6(22): 828. June 1, 1974.

1171. Phillips, James G. "Troubled Coal Industry." NATIONAL JOURNAL REPORTS 6(26): 951-961. June 29, 1974.

1172. Phillipson, John. ECOLOGICAL ENERGETICS. St. Martin's Press, New York. 1966.

1173. Phipps, H. Harry. "Energy Conservation: A New Dimension of Engineering Responsibility." PROFESSIONAL ENGINEER 42: 24-27. November 1972.

1174. Pikl, I. James, Jr. (ed.). PUBLIC POLICY AND THE FUTURE OF THE PETROLEUM INDUSTRY. University of Wyoming, Laramie. 1970.

1175. Pimentel, D. et al. "Food Production and the Energy Crisis." SCIENCE 182( ): 443- . November 2, 1973.

1176. Pinchot, Gifford. "The Long Struggle for Effective Federal Water Power Legislation." GEORGE WASHINGTON LAW REVIEW 14( ): 9-20. December 1945.

1177. Ploch, Louis A. and LeRay, Nelson L. SOCIAL AND ECONOMIC CONSEQUENCES OF THE DICKEY-LINCOLN SCHOOL HYDRO-ELECTRIC POWER DEVELOPMENT ON THE UPPER ST. JOHN VALLEY, MAINE: PHASE I, PRECONSTRUCTION. Miscellaneous Report No. 123. Maine Agricultural Experiment Station, University of Maine, Orono, Maine. March 1968.

1178. Plotnick, Alan R. PETROLEUM: CANADIAN MARKETS AND UNITED STATES FOREIGN TRADE POLICY. University of Washington Press, Seattle. 1965.

1179. Polach, Jaroslav G. "Nuclear Power in Europe at the Crossroads." BULLETIN OF THE ATOMIC SCIENTISTS 25: 15-18, 20. October 1969.

1180. Polach, Jaroslav G. "The Development of Energy in East Europe." In ECONOMIC DEVELOPMENTS IN COUNTRIES OF EASTERN EUROPE: A COMPENDIUM OF PAPERS, Subcommittee on Foreign Policy. Joint Economic Committee. U.S. Congress. U.S. Government Printing Office, Washington, D.C. 1970. (pp. 348-433).

1181. Polzin, Paul E. "Projections of Economic Development Associated with Coal-Related Activity in Montana." Bureau of Business and Economic Research, University of Montana, Missoula. January 1974.

1182. Porter, W.W. "Energy in America." BULLETIN OF THE ATOMIC SCIENTISTS: February 1973.

1183. Posner, Michael V. FUEL POLICY: A STUDY IN APPLIED ECONOMICS. Macmillan London. 1973.

1184. Post, Richard F. "Fusion Power, the Uncertain Certainty." BULLETIN OF THE ATOMIC SCIENTISTS 27( ): 42-48. October 1971.

1185. Powers, Phillip N. "U.S. Must Accept Some Combination of Unattractive Trends in Energy." UNIVERSITY AND ARGONNE RESEARCH ON ENERGY, Argonne University Association Reports. 1973.

1186. Price, Truman. HYDROELECTRIC POWER POLICY. National Technical Information Service, Springfield, Virginia. February 1971.

1187. Priest, A.J. Gustin. PRINCIPLES OF PUBLIC UTILITY REGULATION: THEORY AND APPLICATION. Michie, Charlottesville, Virginia. 1969.

1188. PROFESSIONAL ENGINEER. "Energy in the 1970's." PROFESSIONAL ENGINEER: February 1971.

1189. Pryde, Philip R. and Pryde, Lucy. "Soviet Nuclear Power." ENVIRONMENT 16(3): 26-34. April 1974.

1190. Public Land Law Review Commission. ONE THIRD OF THE NATION'S LAND. Public Land Law Review Commission, Washington, D.C. 1970.

1191. PUBLIC POWER. "Power and the Environment." PUBLIC POWER: May 1970. (Whole Issue).

1192. Pushkarev, Boris S. "Energy in the New York Region." PROCEEDINGS OF THE ACADEMY OF POLITICAL SCIENCE 31(2): 13-23. December 1973.

1193. Putman, Palmer Cosslett. ENERGY IN THE FUTURE. Van Nostrand, New York. 1953.

1194. Queen's University. Canadian Institute of Guided Ground Transport. RAILWAY TO THE ARCTIC: A STUDY OF THE OPERATIONAL AND ECONOMIC FEASIBLITY OF A RAILWAY TO MOVE ARCTIC SLOPE OIL TO MARKET. Summary Report. Queen's University, Kingston, Ontario. 1972. (Revised Edition).

1195. QUEST. "Energy Sources and Their Environmental Effects." QUEST 10(4): 10-13. 1973.

1196. QUEST. "The Role of Government in the Energy Crisis." QUEST 10(4): 4-5. 1973.

1197. Raciti, Sebastian. THE OIL IMPORT PROBLEM. Fordham University Press, New York. 1958.

1198. Radin, Alex. "The Strip Mining Issue." PUBLIC POWER 30(3): 8-10. May-June 1972.

1199. Ramain, Patrice. "Conference Report: French National Symposium on Energy Policy - Paris." ENERGY POLICY 2(1): 74. March 1974.

1200. Ramey, James T. "Delays in Nuclear Plant Licensing: Causes and Possible Solutions." PUBLIC UTILITIES FORTNIGHTLY 89: 19-24. March 30, 1972.

1201. Ramey, James T. "Energy Needs of the Nation and the Cost in Terms of Pollution." ATOMIC ENERGY LAW JOURNAL 14: 26- . Spring 1972.

1202. Ramey, James T. "The Promise of Nuclear Energy." ANNALS OF THE AMERICAN ACADEMY OF POLITICAL AND SOCIAL SCIENCE 410( ): 11-23. November 1973.

1203. Ramey, James T. and Malsch, M.G. "Environmental Quality and the Need for Electric Power -- Legislative Reforms to Improve the Balancing Process." NOTRE DAME LAW REVIEW 47( ): 1139- . June 1972.

1204. Ramey, James T. and Murray, James P., Jr. "Delays and Bottlenecks in the Licensing Process Affecting Utilities: The Role of Improved Procedures and Advance Planning." DUKE LAW JOURNAL 1970: 25-44. February 1970.

1205. Ramseier, Rene O. "Oil on Ice." ENVIRONMENT 16(4): 6-13. April 1974.

1206. Rand Corporation. THE GROWING DEMAND FOR ENERGY. National Technical Information Service, Springfield, Virginia. 1972.

1207. Rand Corporation. ENERGY: A BIBLIOGRAPHY OF SELECTED RAND PUBLICATIONS. Rand Corporation, Santa Monica, California. May 1974.

1208. Randolph, Jennings. "Energy and the Environment; National Fuels and Energy Policy is Needed for U.S.A." NATURAL RESOURCES LAWYER 4: 750-758. November 1971.

1209. Rapoport, Roger. "Catch 24,400 (or, Plutonium is my Favorite Element)." RAMPARTS 8: 16-21. May 1970.

1210. Rappaport, Roy A. PIGS FOR THE ANCESTORS. Yale University Press, New Haven, Connecticut. 1967.

1211. Rappaport, Roy A. "The Flow of Energy in an Agricultural Society." SCIENTIFIC AMERICAN 224(3): 116-133. September 1971.

1212. Rau, Hans. SOLAR ENERGY. Macmillan, New York. 1964.

1213. Ray, Dixy Lee. "Dixy Lee Ray, Chief of the Atomic Energy Commission: An Interview." ENVIRONMENTAL QUALITY 4: 19-26. July 1973.

1214. Ray, Dixy Lee et al. THE NATION'S ENERGY FUTURE. Report to the President of the United States. U.S. Government Printing Office, Washington, D.C. 1973.

1215. Reed, John C. "Effects of Oil Development in Arctic America." BIOLOGICAL CONSERVATION 2: 273-277. July 1970.

1216. Refi, William E. and Knoke, William A. "The Marketing Concept for Combination Gas and Electric Companies." PUBLIC UTILITIES FORTNIGHTLY 86: 33-69. October 22, 1970.

1217. Reichle, Leonard F.C. "Nuclear Power -- 1970-80." PUBLIC UTILITIES FORTNIGHTLY 85: 26-43. February 12, 1970.

1218. Reid, William T. "What About Air Pollution by Power Plants?" BATTELLE RESEARCH OUTLOOK 2(3): 21-24. 1970.

1219. Reifsnyder, C. Frank. "Natural Gas Pipelines and the Courts." NATURAL RESOURCES LAWYER 2: 221-228. July 1969.

1220. Reinemer, Vic. "The Relation of Power Needs to Population Growth, Economic and Social Costs." In ENERGY, THE ENVIRONMENT, AND HUMAN HEALTH, Asher J. Finkel (ed.). Publishing Sciences Group, Acton, Massachusetts. 1973.

1221. Reis, Robert I. "Environmental Activism: Thermal Pollution -- AEC and State Jurisdictional Consideration." BOSTON COLLEGE INDUSTRIAL AND COMMERCIAL LAW REVIEW 13( ): 633- . March 1972.

1222. Renshaw, Edward F. "Substitution of Inanimate Energy for Animal Power." JOURNAL OF POLITICAL ECONOMY 71: 284-292. June 1963.

1223. Resources for the Future. ENERGY, ECONOMIC GROWTH, AND THE ENVIRONMENT. Papers Presented at a forum conducted by Resources for the Future, Inc., April 20-21, 1971, Washington, D.C. Published for Resources for the Future by the Johns Hopkins Press, Baltimore. 1972.

1224. Resources for the Future, Inc. ENERGY RESEARCH NEEDS. National Technical Information Service, Springfield, Virginia. 1972.

1225. Resources for the Future and MIT Environmental Laboratory. ENERGY RESEARCH NEEDS: CONSUMPTION, PRODUCTION, TECHNOLOGY, ENVIRONMENTAL EFFECTS, POLICY ISSUES. Resources for the Future, Washington, D.C. 1971.

1226. Rhode, G.K. "Utility Environmental Surveillance Programs." AMERICAN JOURNAL OF PUBLIC HEALTH: February 1972.

1227. Rhyne, Charles S. et al. CITIES AND ATOMIC ENERGY. Report Number 145. National Institute of Law Officers, Chicago. 1959.

1228. Ribicoff, Abraham. PETROPOLITICS AND THE AMERICAN ENERGY SHORTAGE. Report to the Committee on Government Operations of the U.S. Senate. 93rd Congress. U.S. Government Printing Office, Washington, D.C. 1973.

1229. Rich, Spencer. "The Power of Oil." PROGRESSIVE 33: 19-23. September 1969.

1230. Riddick, Winston W. "The Nature of the Petroleum Industry." PROCEEDINGS OF THE ACADEMY OF POLITICAL SCIENCE 31(2): 148-158. December 1973.

1231. Ridgeway, James. THE LAST PLAY: THE STRUGGLE TO MONOPOLIZE THE WORLD'S ENERGY RESOURCES. E.P. Dutton and Company, New York. 1973.

1232. Ridgeway, James. "Notes on the Energy Crisis." RAMPARTS 12: 37-41. October 1973.

1233. Ripley, Anthony. "Atomic Power Abuse: The AEC in Colorado." WASHINGTON MONTHLY 2: 6-14. July 1970.

1234. Risser, Hubert E. THE ECONOMICS OF THE COAL INDUSTRY. Bureau of Business Research, University of Kansas, Lawrence. 1958.

1235. Risser, Hubert E. COAL IN THE FUTURE ENERGY MARKET. Illinois State Geological Survey, Urbana. 1960.

1236. Risser, Hubert E. EVALUATION OF FUELS -- LONG-TERM FACTORS AND CONSIDERATIONS. Illinois State Geological Survey, Urbana. 1969.

1237. Risser, Hubert E. POWER AND THE ENVIRONMENT: A POTENTIAL CRISIS IN ENERGY SUPPLY. Environmental Geology Notes Number 40. Illinois State Geological Society, Urbana, Illinois. 1970.

1238. Ritchie, Ronald S. "Canada's Energy Situation in a World Context." INTERNATIONAL PERSPECTIVES: 13-17. March-April 1974.

1239. Rivera-Cordero, Antonio. "The Nuclear Industry and Air Pollution." ENVIRONMENTAL SCIENCE AND TECHNOLOGY 4: 392-395. May 1970.

1240. Robbins, Charles et al. "U.S. Atomic Energy Industry." NUCLEAR ENGINEERING INTERNATIONAL 15: 905-929. November 1970.

1241. Roberts, F.S. "Signed Digraphs and the Growing Demand for Energy." ENVIRONMENT AND PLANNING 3(4): 395-410. 1971.

1242. Roberts, Marc J. "Is There an Energy Crisis?" PUBLIC INTEREST (31): 17-37. Spring 1973.

1243. Robie, Edward H. ECONOMICS OF THE MINERAL INDUSTRIES. American Institute of Mining, Metallurgical, and Petroleum Engineers, New York. 1964.

1244. Robinson, Colin and Crook, Elizabeth M. "Is There a World Energy Crisis?" NATIONAL WESTMINSTER BANK QUARTERLY REVIEW: 46. May 1973.

1245. Rocks, Lawrence and Runyon, Richard P. THE ENERGY CRISIS. Crown Publishers, New York. 1972.

1246. Rocky Mountain Petroleum Institute. BALANCING SUPPLY AND DEMAND FOR ENERGY IN THE UNITED STATES. Rock Mountain Petroleum Institute, University of Denver. 1972.

1247. Rogers. "Siting Power Plants in Washington State." WASHINGTON LAW REVIEW 47( ): 9- . 1971.

1248. Rogers, George W. (ed.). CHANGE IN ALASKA, PEOPLE, PETROLEUM, POLITICS. University of Alaska Press, College, Alaska; University of Washington Press, Seattle. 1970.

1249. Rogers, Walter E. "Is There a National Shortage of Gas in Our Future?" PUBLIC UTILITIES FORTNIGHTLY 85: 17-22. March 26, 1970.

1250. Rooney, Robert F. TAXATION AND REGULATION OF THE DOMESTIC OIL AND GAS INDUSTRY. Unpublished Ph.D. dissertation, Stanford University, 1965.

1251. Rose, David J. "Nuclear Fuels (Fission-Fusion)." In ENERGY, THE ENVIRONMENT, AND HUMAN HEALTH, Asher J. Finkel (ed.). Publishing Sciences Group, Acton, Massachusetts. 1973.

1252. Rosenberg, Nathan. "Factors Affecting the Diffusion of Technology." EXPLORATIONS IN ECONOMIC HISTORY 10(1): 3-34. Fall 1972.

1253. Ross, Charles R. "Electricity as a Social Force." THE ANNALS OF THE AMERICAN ACADEMY OF POLITICAL AND SOCIAL SCIENCE 405: 47-54. January 1973.

1254. Ross, John E. and Jenkins, Sarah. "Newspaper Coverage of Nuclear Power Plant Issues: Lake Michigan, 1966-1969." Institute for Environmental Studies, University of Wisconsin, Madison. 1972.

1255. Rouhani, Fuad. A HISTORY OF O.P.E.C. Praeger, New York. 1971.

1256. Rubin, David M. et al. "Public Policy Toward Environment 1973: A Review and Appraisal: Environmental Information." ANNALS OF THE NEW YORK ACADEMY OF SCIENCE 216: 171-177. 1973.

1257. Rubin, David M. and Sachs, David P. MASS MEDIA AND THE ENVIRONMENT: THE PRESS DISCOVERS THE ENVIRONMENT. Volume II. Educational Resources Information Center, Bethesda, Maryland. 1971.

1258. Rubin, David M. and Sachs, David P. (eds.). MASS MEDIA AND THE ENVIRONMENT: WATER RESOURCES, LAND USE AND ATOMIC ENERGY IN CALIFORNIA. Praeger, New York. 1973.

1259. Ruckelshaus, William D. "Energy and Environment on a Collision Course." ECOLOGY TODAY 2: 2-4. March-April 1972.

1260. Runyon, Richard P. and Rocks, Lawrence. "Energy and Foreign Policy: How Dependent Must We Be?" In ENERGY AND THE ENVIRONMENT: A COLLISION OF CRISES, Irwin Goodwin (ed.). Publishing Sciences Group, Acton, Massachusetts. 1973.

1261. Runyon, Richard P. and Rocks, Lawrence. "The Energy Crisis." PROCEEDINGS OF THE ACADEMY OF POLITICAL SCIENCE 31(2): 3-12. December 1973.

1262. Russell, Clifford S. RESIDUALS MANAGEMENT IN INDUSTRY: A CASE STUDY OF PETROLEUM REFINING. The Johns Hopkins University Press, Baltimore. 1973.

1263. Russell, Milton and Toenjes, Lawrence. NATURAL GAS PRODUCER REGULATION AND TAXATION, Institute of Public Utilities, Michigan State University, East Lansing, Michigan. 1971.

1264. RUTGERS LAW REVIEW. "The Regulation of Nuclear Power After the National Environmental Policy Act of 1969." RUTGERS LAW REVIEW 24: 753-769. Summer 1970.

1265. Ruttenberg, Stanley H. and Associates. THE ELECTRIC POWER CRISIS: ITS IMPACT ON WORKERS AND CONSUMERS. National Rural Electric Cooperative Association, Washington, D.C. 1971.

1266. Ryan, John J. and Welles, John G. REGIONAL ECONOMIC IMPACT OF A U.S. OIL SHALE INDUSTRY. Denver Research Institute, Denver University, Denver. 1966.

1267. Sabate, Robert W. "Big, Bad Oil: Not So Bad After All." AMERICAN BAR ASSOCIATION JOURNAL 60: 716- . June 1974.

1268. Sage, Bryan L. "Oil and Alaskan Ecology." NEW SCIENTIST 46: 175-177. April 1970.

1269. Sagan, L.A. "Human Costs of Nuclear Power." SCIENCE 177(4048): 487-493. August 11, 1972.

1270. Sailor, Vance L. "The United States Energy Crisis: Supply and Demand." Paper Presented at the Annual Conference of the Science Teachers Association of New York State, Kiamesha Lake, New York. October 1972.

1271. Sakisaka, Masao. "World Energy Problems and Japan's International Role." ENERGY POLICY 1(2): 100-106. September 1973.

1272. Santacruz, Hernan. "The Energy Crisis." REVIEW OF INTERNATIONAL AFFAIRS 25(575): 1-3. March 20, 1974.

1273. Sargent, Francis W. "The Need for a New Perspective on Energy." PROCEEDINGS OF THE ACADEMY OF POLITICAL SCIENCE 31(2): 24-32. December 1973.

1274. Saskatchewan. Provincial Library, Regina. Bibliographic Services Division. ENERGY CRISIS: A BIBLIOGRAPHY. Provincial Library, Regina, Saskatchewan. 1973.

1275. Saunders, William T. "Energy Flow in Neolithic Ecosystems." Paper Presented at the Annual Meeting of the American Association for the Advancement of Science, San Francisco. February 1974.

1276. Savery, C. William. "Future Energy Sources for Transportation." TRAFFIC QUARTERLY 26(4): 485-500. October 1972.

1277. Sax, Joseph L. DEFENDING THE ENVIRONMENT. Knopf, New York. 1971.

1278. Scargill, D.I. "Energy in France." GEOGRAPHY 58(259): 159-162. April 1973.

1279. Schatz, Joel. "Cosmic Economics." In ENERGY: TODAY'S CHOICES, TOMORROW'S OPPORTUNITIES, Anton B. Schmalz (ed.). World Future Society, Washington, D.C. 1974. (pp. 20-26).

1280. Schiele, Robin. "Inflation and Energy Crises, Czech Style." CANADIAN BUSINESS 47(6): 51-52. June 1974.

1281. Schlesinger, James R. "Energy, the Environment and Society." ATOMIC ENERGY LAW JOURNAL 14: 3. Spring 1972.

1282. Schmalz, Anton B. (ed.). ENERGY: TODAY'S CHOICES, TOMORROW'S OPPORTUNITIES: ESSENTIAL DIMENSIONS IN THINKING FOR ENERGY POLICY. World Future Society, Washington, D.C. 1974.

1283. Schmandt, Jurgen. ONE ASPECT OF THE ENERGY CRISIS: THE UNBALANCED STATE OF ENERGY R&D. Research Report Number 1. Occasional Papers. Lyndon B. Johnson School of Public Affairs, Austin, Texas. 1972.

1284. Schmidt, Herman J. "The Government's Role in Energy." THE CONFERENCE BOARD RECORD 11(5): 20-22. May 1974.

1285. Schmidt-Bleek, F. and Carlsmith, R.S. (eds.). SYMPOSIUM ON COAL AND PUBLIC POLICIES, HELD AT THE UNIVERSITY OF TENNESSEE, KNOXVILLE, OCTOBER 13-15, 1971. Center for Business and Economic Research, College of Business Administration, University of Tennessee, Knoxville. 1972.

1286. Schoenbrod, D. and Case, C.P., III. "Electricity or the Environment: A Study of Public Regulation Without Public Control." CALIFORNIA LAW REVIEW 61( ): 961. 1973.

1287. Schoener, Thomas W. "Population Growth Regulated by Intraspecific Competition for Energy or Time: Some Simple Representations." THEORETICAL POPULATION BIOLOGY 4(1): 56-84. March 1973.

1288. Schooler, Dean. "The Politics of Environment and National Development." In GROWING AGAINST OURSELVES: THE ENERGY-ENVIRONMENT TANGLE, S.L. Kwee and J.S.R. Mullender (eds). Lexington Books, Lexington, Massachusetts. 1972.

1289. Schrader, Gene. "Atomic Doubletalk." CENTER MAGAZINE 4: 29-52. January-February 1971.

1290. Schurr, Sam H. SOME OBSERVATIONS ON THE ECONOMICS OF ATOMIC POWER. Resources for the Future, Washington, D.C. 1963.

1291. Schurr, Sam H. "Some Aspects of the Evolution of Petroleum Policies in the United States." In PUBLIC POLICY AND THE FUTURE OF THE PETROLEUM INDUSTRY, I.J. Pikl (ed.). University of Wyoming, Laramie. 1970.

1292. Schurr, Sam H. ENERGY RESEARCH NEEDS. Published by Resources for the Future by The Johns Hopkins Press, Baltimore. 1971.

1293. Schurr, Sam H. (ed.). ENERGY, ECONOMIC GROWTH AND THE ENVIRONMENT. Published for Resources for the Future by The Johns Hopkins Press, Baltimore. 1972.

1294. Schurr, Sam H., et al. ENERGY IN THE AMERICAN ECONOMY, 1850-1975: ITS HISTORY AND PROSPECTS. The Johns Hopkins Press, Baltimore. 1960.

1295. Schurr, Sam H., et al. MIDDLE EASTERN OIL AND THE WESTERN WORLD: PROSPECTS AND PROBLEMS. American Elsevier, New York. 1971.

1296. Schurr, Sam H. and Eliasberg, Vera B. ENERGY AND ECONOMIC GROWTH IN THE UNITED STATES. Resources for the Future, Washington, D.C. 1962.

1297. Schurr, Sam H. and Marschak, Jacob. ECONOMIC ASPECTS OF ATOMIC POWER. Princeton University Press, Princeton, New Jersey. 1950.

1298. Schnaiberg, Allan. "The First and Last Dialectic: Social Impacts of Environmental Quality and Societal Scarcity." Paper Presented at the Annual Meeting of the Society for the Study of Social Problems, Montreal. August 1974.

1299. Schwartz, T.P. and Schwartz-Barcott, Donna. "The Short End of the Shortage: On The Self-Reported Impact of the Energy Shortage on the Socially Disadvantaged." Paper Presented at the Annual Meeting of the Society for the Study of Social Problems, Montreal. August 1974.

1300. SCIENCE (eds.). "SCIENCE Bibliography of Energy." SCIENCE 184 (4134): 386-388. April 19, 1974.

1301. SCIENCE NEWS. "Judging the Energy Crisis." SCIENCE NEWS: November 14, 1970.

1302. SCIENCE NEWS. "Energy, Technology and Future Options." SCIENCE NEWS: July 24, 1971.

1303. SCIENTIFIC AMERICAN (eds.). ENERGY AND POWER: READINGS FROM SCIENTIFIC AMERICAN. Freeman, San Francisco. 1971.

1304. SCIENTIFIC AMERICAN (eds.). "Energy." SCIENTIFIC AMERICAN 184 (4134). April 1974. (Whole Issue).

1305. Scott, David. POLLUTION IN THE ELECTRIC POWER INDUSTRY: ITS CONTROL AND COSTS. Lexington Books, Lexington, Massachusetts. 1973.

1306. Scott, R. Stephen. "A Look at the Nuclear Energy Option." MAN/SOCIETY/TECHNOLOGY 33(1): 7-12. September-October 1973.

1307. Seaborg, Glenn T. "Looking Ahead in Nuclear Power." EDISON ELECTRIC INSTITUTE BULLETIN 37: 188-195, 231. June-July 1969.

1308. Seaborg, Glenn T. "The Nuclear Plant and Our Energy Needs." PUBLIC UTILITIES FORTNIGHTLY 85: 19-26. February 12, 1970.

1309. Seaborg, Glenn T. "Energy and Environment." THE INTERNATIONAL JOURNAL OF ENVIRONMENTAL STUDIES 3(4): 301-306. September 1972.

1310. Seaborg, Glenn T. "Energy and Our Future." PUBLIC UTILITIES FORTNIGHTLY 91: 13-17. February 1, 1973.

1311. Seaborg, Glenn T. and Corliss, William. MAN AND ATOM. Dutton, New York. 1971.

1312. Searl, Milton. "The Supply of Fuels as a Function of Price." Paper Presented at a meeting of the American Association for the Advancement of Science, San Francisco, California. February 1974.

1313. Seidel, Marquis R., Plotkin, Steven E. and Reck, Robert O. ENERGY CONSERVATION STRATEGIES. U.S. Government Printing Office, Washington, D.C. 1973.

1314. Seifert, William W., Bakr, Mohammet A. and Kettani, M. Ali.
ENERGY DEVELOPMENT, A CASE STUDY. M.I.T. Press, Cambridge,
Massachusetts. 1973.

1315. Sell, George. THE PETROLEUM INDUSTRY. Oxford University Press,
New York. 1963.

1316. Seulflow, James E. "The FPC Hydro License Controversy." PUBLIC
UTILITIES FORTNIGHTLY 84: 28-30. July 31, 1969.

1317. Sewell, W.R. Derrick. "Alaska Pipeline Regardless of Cost."
GEOGRAPHICAL MAGAZINE 46(8): 383-386. May 1974.

1318. Shaffer, Edward H. THE OIL IMPORT PROGRAM OF THE UNITED STATES:
AN EVALUATION. Praeger, New York. 1968.

1319. Shapley, D. "Law of the Sea: Energy, Economy Spur Secret Review
of U.S. Stance." SCIENCE 183( ): 290. 1974.

1320. Shaw, Milton. "Is Atomic Power The Answer?" In ENERGY AND THE
ENVIRONMENT: A COLLISION OF CRISES, Irwin Goodwin (ed.).
Publishing Sciences Group, Acton, Massachusetts. 1973.

1321. Shell Oil Company, Inc. SHELL ASSESSES THE NATIONAL FUEL SUPPLY
PROBLEM. Shell Oil Company, New York. 1971.

1322. Shepherd, William G. and Gies, Thomas G. UTILITY REGULATION: NEW
DIRECTIONS IN THEORY AND POLICY. Random House, New York. 1966.

1323. Sherrill, Robert. "Energy Crises! The Industry's Fright Campaign."
NATION 214(26): 816-820. June 26, 1972.

1324. Shoupp, W.E. WORLD ENERGY AND THE OCEANS. Second Annual Sea
Grant Lecture and Symposium. Report Number MITSG 74-7.
Massachusetts Institute of Technology, Cambridge, Massachusetts.
October 1973.

1325. Shumway, F.R. "Energy Crisis." Address December 10, 1970. VITAL
SPEECHES. January 15, 1971.

1326. Shutler, N.D. "Pollution of the Sea by Oil." HOUSTON LAW REVIEW
7( ): 415-441. March 1970.

1327. Shuttleworth, John, et al. THE MOTHER EARTH NEWS HANDBOOK OF HOME-
MADE POWER. Bantam Books, New York. 1974.

1328. Shwadran, Benjamin. THE MIDDLE EAST, OIL AND THE GREAT POWERS.
Halstead Press, New York. 1974.

1329. Sien-chong, Niu. "China's Petroleum Industry." MILITARY REVIEW
49: 23-27. November 1969.

1330. Sillin, Lelan F., Jr. "A Commitment to Change--Utilities, Regulators, and Conservationists." PUBLIC UTILITIES FORTNIGHTLY 85: 48-54. June 4, 1970.

1331. Sills, David L. "The Environmental Movement and Its Critics." Paper Presented at the Annual Meeting of the American Association for the Advancement of Science, San Francisco. February 1974.

1332. Simon, Jack A. "The Impending Energy Crisis: A Look at the Coming Decade." ILLINOIS BUSINESS REVIEW. June 1972.

1333. Simon, William E. "Energy in the USA After the President's Message. . . Changes in Investment and Balance of Payments." ENERGY POLICY 1(3): 187-194. December 1973.

1334. Simpson, John W. "Perspectives on Power and Policy." PUBLIC UTILITIES FORTNIGHTLY 91: 34-39. March 29, 1973.

1335. Simpson, R.D.H. "Further Remarks on the Future of Oil." THE GEOGRAPHICAL JOURNAL 139(3): 455-459. October 1973.

1336. Singer, S. Fred. "Human Energy Production as a Process in the Biosphere." SCIENTIFIC AMERICAN 222(3): 175-190. September 1970.

1337. Sinnott, Edmund W. "Man and Energy." YALE REVIEW 38: 640-653. June 1949.

1338. Siri, William E. "Power Plant Sitings: View of Environmentalists." In ENERGY, THE ENVIRONMENT, AND HUMAN HEALTH, Asher J. Finkel (ed.). Publishing Sciences Group, Acton, Massachusetts. 1973.

1339. Sireta, H. and Schwartz, M. "Agriculture and Energy." ORGANIC GARDENING AND FARMING. September 1973.

1340. Sive. "The Role of Litigation in Environmental Policy: The Power Plant Siting Problem." NATURAL RESOURCES JOURNAL 11( ): 467. 1971.

1341. Skeet, Trevor. "Oil in Africa." AFRICAN AFFAIRS 70: 72-76. January 1971.

1342. Slater, N. "The Changing Posture of the Nuclear Community." PROFESSIONAL ENGINEER. February 1971.

1343. Smernoff, Barry J. "Energy Policy Interactions in the United States." ENERGY POLICY 1(2): 136-153. September 1973.

1344. Smil, Vaclav. "Energy and the Environment--A Delphi Forecast." LONG RANGE PLANNING 5(4): 27-32. December 1972.

1345. Smil, Vaclav. "Energy and the Environment." THE FUTURIST 8(1): 4-13. February 1974.

1346. Smith, Dan. "The Phony Oil Crisis: A Survey." ECONOMIST 248: 1-42. July 7, 1973.

1347. Smith, David. "Inter-Energy Competition Between Gas and Electricity." ARIZONA BUSINESS. October 1972.

1348. Smith, Frank E. (ed.). CONSERVATION IN THE UNITED STATES, A DOCUMENTED HISTORY: LAND AND WATER 1900-1970. Chelsea House, New York. 1971.

1349. Smith, Fred. "The Electric Utility Task Force Report--One Year Later." EDISON ELECTRIC INSTITUTE BULLETIN 37: 317-321, 339. October 1969.

1350. Smith, J.E. TORREY CANYON, POLLUTION AND MARINE LIFE. U.K. Marine Biological Association, Cambridge University Press, Cambridge, Massachusetts. 1968.

1351. Smith, J.N. "Influence of State and local Government on Energy Demand." In STATE AND LOCAL DECISION-MAKING ON ENERGY POLICY, The Energy Policy Project, Washington, D.C. 1973.

1352. Smith, Robert. "A Regulator's View on Rate Control." PUBLIC UTILITIES FORTNIGHTLY 84: 26-34. September 25, 1969.

1353. Smith, Robert. "Canadian Gas Export Policy." PUBLIC UTILITIES FORTNIGHTLY 86: 23-27. November 5, 1970.

1354. Smith, Stephen. "The Energy Crisis: Reality or Myth?" ANNALS OF THE AMERICAN ACADEMY OF POLITICAL AND SOCIAL SCIENCE: November 1973.

1355. Smith, Turner T., Jr. "Electricity and The Environment--The Generating Plant Siting Problem." BUSINESS LAWYER 26: 169-197. November 1970.

1356. Smith, William D. "Shortage Amid Plenty." PROCEEDINGS OF THE ACADEMY OF POLITICAL SCIENCE 31(2): 41-52. December 1973.

1357. Snow, Donald L. "Standards Needs in Controlling Radiation Exposure of the Public." AMERICAN JOURNAL OF PUBLIC HEALTH AND THE NATION'S HEALTH 60: 243-349. February 1970.

1358. Snyder, M.J. and Chilton, Cecil. "Planning for Uncertainty: Energy in the Years 1975-2000." BATTELLE RESEARCH OUTLOOK 4(1). 1972.

1359. Society of Economic Geologists. THE MINERAL POSITION OF THE UNITED STATES, 1975-2000. University of Wisconsin Press, Madison. 1974.

1360. South Dakota Planning Bureau. "Some Impacts on South Dakota of Coal-Related Development in the Northern Great Plains." State Planning Bureau, Pierre, South Dakota. April 1974.

1361. Spaak, Fernand. "An Energy Policy for the European Community." ENERGY POLICY 1(1): 35-37. June 1973.

1362. Spangler, Miller B. "Environmental and Social Issues of Site Choice for Nuclear Power Plants." ENERGY POLICY 2(1): 18-32. March 1974.

1363. Sparling, Richard C., et al. FINANCIAL ANALYSIS OF THE PETROLEUM INDUSTRY. Chase Manhattan Bank, New York. 1971.

1364. Sparling, Richard C., Anderson, Norma J. and Winger, John G. CAPITAL INVESTMENT OF THE WORLD PETROLEUM INDUSTRY--1968. Chase Manhattan Bank, New York. 1969.

1365. Spero, Joan Edelman. "Energy Self-Sufficiency and National Security." PROCEEDINGS OF THE ACADEMY OF POLITICAL SCIENCE 31(2): 123-136. December 1973.

1366. Sperry, K. "North Cascades National Park: Copper Mining vs. Conservation." SCIENCE 157: 1021-1024. September 1, 1967.

1367. Sporn, Philip. "Observations on Private Versus Public Power." PUBLIC UTILITIES FORTNIGHTLY 53: 717-733. June 10, 1954.

1368. Sporn, Philip. ENERGY: ITS PRODUCTION, CONVERSION AND USE IN THE SERVICE OF MAN. Macmillan, New York. 1965.

1369. Sporn, Philip. NUCLEAR POWER ECONOMICS, 1962-1967. U.S. Government Printing Office, Washington, D.C. 1968.

1370. Sporn, Philip. THE SOCIAL ORGANIZATION OF ELECTRIC POWER SUPPLY IN MODERN SOCIETIES. M.I.T. Press, Cambridge, Massachusetts. 1971.

1371. Sporn, Philip. "The Indispensability of a National Energy Policy. . . And of Research to Avert an Energy Crisis." MANAGEMENT QUARTERLY 14(2): 2-9. Summer 1973.

1372. Sporn, Philip. "Multiple Failures of Public and Private Institutions." SCIENCE 184(4134): 284-286. April 19, 1974.

1373. Sprout, Harold and Sprout, Margaret. "The Dilemma of Rising Demands and Insufficient Resources." WORLD POLITICS 20(4): 660-693. July 1968.

1374. Sprout, Harold and Sprout, Margaret. TOWARD A POLITICS OF THE PLANET EARTH. Van Nostrand/Reinhold, New York. 1971.

1375. Sprout, Harold and Sprout, Margaret. "National Priorities: Demands, Resources, Dilemmas." WORLD POLITICS 24: 293. 1972.

1376. Sprout, Harold and Sprout, Margaret. "Public Policy and Environmental Crisis." PUBLIC POLICY JOURNAL 1: 192. 1973.

1377. Sprout, Harold and Sprout, Margaret. MULTIPLE VULNERABILITIES: THE CONTEXT OF ENVIRONMENTAL REPAIR AND PROTECTION. Research Monograph Number 40. Center of International Studies, Woodrow Wilson School of Public and International Studies, Princeton University, Princeton. April 1974.

1378. Stacks, John F. STRIPPING. Sierra Club, New York. 1972.

1379. Stanford Research Institute. END USES OF ENERGY. Stanford Research Institute, Menlo Park, California. 1971.

1380. Stanford Research Institute. PATTERNS OF ENERGY CONSUMPTION IN THE UNITED STATES. Report Prepared for the Office of Science and Technology, Executive Office of the President. Stanford Research Institute, Menlo Park, California. 1972.

1381. Stanford Research Institute. Long Range Planning Service. THERMOELECTRIC POWER. Long Range Planning Report Number 55, Stanford Research Institute, Menlo Park, California. 1960.

1382. Stanford Research Institute. Long Range Planning Service. THE GROWING NEED FOR ENERGY. Long Range Planning Report Number 52, Stanford Research Institute, Menlo Park, California. 1962.

1383. Stanford Research Institute. Long Range Planning Service. INTER-FUEL COMPETITION IN THE UNITED STATES. Stanford Research Institute, Menlo Park, California. 1967.

1384. Stanley, Manfred. "Energy as Metaphor." Paper Presented at the Annual Meeting of the American Association for the Advancement of Science, San Francisco. February 1974.

1385. Starr, Chauncey. "National Atomic Energy Policy--Time For a Change." PROFESSIONAL ENGINEER. February 1971.

1386. Starr, Chauncey. "Energy and Power." SCIENTIFIC AMERICAN 225: 25, 36-49. September 1971.

1387. Starr, Chauncey. "Research on Future Energy Systems." In EMERGENCY POLICY AND NATIONAL GOALS. Hearings Before Senate Committee on Interior and Insular Affairs. October 20, 1971. Part II. U.S. Government Printing Office, Washington, D.C. 1971.

1388. Starr, Chauncey. "Energy R & D Planning." SCIENCE POLICY REVIEWS (3). 1972.

1389. Starr, Chauncey. "Technological Potentials for Energy Development." THE CONFERENCE BOARD RECORD 9(7): 22. July 1972.

1390. Starr, Chauncey. "International Realities of Energy Crisis." DEPARTMENT OF STATE NEWSLETTER (141): 20-25. January 1973.

1391. Starr, Chauncey. "Expanded R & D for the Electric Utilities." EEI BULLETIN 41: 120-122, 132. May-June 1973.

1392. Starr, Chauncey. "Realities of the Energy Crisis." SCIENCE AND PUBLIC AFFAIRS BULLETIN OF THE ATOMIC SCIENTISTS 29(7): 15-20. September 1973.

1393. Starr, Robert. "Power and The People--The Case of Con Edison." THE PUBLIC INTEREST 26: 75-99. Winter 1972.

1394. Stauffer, Robert F. "President Nixon's Energy Message and the Coal Industry Lawyer." NATURAL RESOURCES LAWYER 6(4): 543-552. Fall 1973.

1395. Stauffer, Thomas R. "Economic Cost of U.S. Crude Oil Production." JOURNAL OF PETROLEUM TECHNOLOGY 25: 643-658. June 1973.

1396. Stauffer, Thomas R. "Oil Money and World Money: Conflict or Confluence?" SCIENCE 184(4134): 321-324. April 19, 1974.

1397. Stauffer, Thomas R. and Jensen, James T. "The Rational Allocation of Natural Gas Under Chronic Supply Constraints." In ENERGY: DEMAND, CONSERVATION AND INSTITUTIONAL PROBLEMS, Michael Macrakis (ed.). M.I.T. Press, Cambridge, Massachusetts. 1974.

1398. Stead, William J. "The Sun and Foreign Policy." BULLETIN OF THE ATOMIC SCIENTISTS: 86-90. March 1957.

1399. Steele, Henry B. and Rustow, Dankwart A. OIL IMPORTS AND THE NATIONAL INTEREST. Petroleum Industry Research Foundation, New York. 1971.

1400. Stein, Richard G. "Spotlight on the Energy Crisis: How Architects Can Help." AIA JOURNAL 57(6): 18-23. June 1972.

1401. Stein, Richard G. "A Matter of Design." ENVIRONMENT 14(8): 16-29. October 1972.

1402. Stein, Richard G. "Architecture and Energy." ARCHITECTURAL FORUM: July/August 1973.

1403. Steinhart, Carol E. and Steinhart, John S. BLOWOUT: A CASE STUDY OF THE SANTA BARBARA OIL SPILL. Duxbury, Belmont, California. 1972.

1404. Steinhart, Carol E. and Steinhart, John S. ENERGY: SOURCES, USE AND ROLE IN HUMAN AFFAIRS. Duxbury, Belmont, California. 1973.

1405. Steinhart, John S. and Steinhart, Carol E. "Energy Use in the U.S. Food System." SCIENCE 184(4134): 307-315. April 19, 1974.

1406. Stelzer, Irwin M. and Frazier, Charles J. "Industry Economics-- A Time for Realism by Gas Men." PUBLIC UTILITIES FORTNIGHTLY 88: 29-34. December 23, 1971.

1407. Stephenson, Charles M. "Implications of PLLRC Tax Recommendations for Federal Hydro Projects and Power Facilities." LAND ECONOMICS. February 1973.

1408. Stern, Carlos. "Hydro-Power Vs. Wilderness Waterway: A Case Study of A Federal Water Resources Conflict on the Upper Missouri." Department of Agricultural Economics, University of Connecticut, Storrs. 1973.

1409. Stewart, Charles F. "Energy and the Balance of Payments." PROCEEDINGS OF THE ACADEMY OF POLITICAL SCIENCE 31(2): 137-147. December 1973.

1410. Stich, Robert S. "Financial Aspects of German Electric Utilities." PUBLIC UTILITIES FORTNIGHTLY 84: 27-31. August 28, 1969.

1411. Stipak, Brian. "An Analysis of the Rapid Transit Vote in Los Angeles." TRANSPORTATION 2(1): 71- . April 1973.

1412. Stocking, George W. MIDDLE EAST OIL: A STUDY IN POLITICAL AND ECONOMIC CONTROVERSY. Vanderbilt University Press, Nashville, Tennessee. 1970.

1413. Stoltenberg, Carl H., et al. PLANNING RESEARCH FOR RESOURCE DECISIONS. Iowa State University Press, Ames, Iowa. 1971.

1414. Stone, Lewis Bart. "Power Siting: A Challenge to the Legal Process." ALBANY LAW REVIEW 36: 1-34. Fall 1971.

1415. Stout, B.A. "Energy in Agriculture and Natural Resources: Selected References." Compiled for Energy Task Force. Department of Agricultural Engineering, Michigan State University, East Lansing. August 1974.

1416. Strout, Alan M. TECHNOLOGICAL CHANGE AND UNITED STATES ENERGY CONSUMPTION 1939-1954. Unpublished Ph.D. dissertation, University of Chicago. 1966.

1417. Study of Critical Environmental Problems. Work Group on Energy Products. MAN'S IMPACT ON THE GLOBAL ENVIRONMENT: ASSESSMENT AND RECOMMENDATIONS FOR ACTION. M.I.T. Press, Cambridge, Massachusetts. 1970.

1418. Stunkel, K.R. "The Technological Solution." SCIENCE AND PUBLIC AFFAIRS :42. September 1973.

1419. Surrey, A.J. "The Future Growth of Nuclear Power: Part I. Demand and Supply." ENERGY POLICY 1(2): 107-129. September 1973.

1420. Surrey, A.J. "The Future Growth of Nuclear Power: Part II. Choices and Obstacles." ENERGY POLICY 1(3): 208-224. December 1973.

1421. Surrey, A.J. and Bromley, A.J. "Energy Resources." In MODELS OF DOOM: A CRITIQUE OF THE LIMITS TO GROWTH, H.S.D. Cole, et al. (eds.). Universe Books, New York. 1973. (pp. 90-107).

1422. Surrey, A.J. and Bromley, A.J. "Energy Resources." FUTURES 5(1): 90-107. February 1973.

1423. Surrey, A.J. and Chesshire, J.H. WORLD MARKET FOR ELECTRIC POWER EQUIPMENT: RATIONALIZATION AND TECHNICAL CHANGE. Science and Policy Research Unit, University of Sussex, Brighton. 1972.

1424. Surrey, John. "The Manufacture of Heavy Electrical Plants in Western Europe--Problems and Prospects." INDUSTRIAL MARKETING MANAGEMENT 1(4): 449-457. July 1972.

1425. SURVEY OF CURRENT AFFAIRS. "Energy." SURVEY OF CURRENT AFFAIRS 3(11): 457-461. November 1973.

1426. Swan, Peter N. "International and National Approaches to Oil Pollution Responsibility: An Emerging Regime for a Global Problem." OREGON LAW REVIEW 50( ): 506-586. 1971.

1427. Swanson, Edward B. A CENTURY OF OIL AND GAS IN BOOKS, A DESCRIPTIVE BIBLIOGRAPHY. Appleton-Century-Crofts, New York. 1960.

1428. Sweeney, Joseph C. "Oil Pollution of the Oceans." FORDHAM LAW REVIEW 37( ): 155-208. 1971.

1429. Swengel, F.M. "A New Era of Power Supply Economics." POWER ENGINEERING 74: 30-38. March 1970.

1430. Swidler, Joseph C. "The Critical Path Between Energy and the Environment." PUBLIC POWER 29: 18-20, 48. November-December 1971.

1431. Swidler, Joseph C. "The Role of Energy Conservation in a National Energy Policy." ENVIRONMENTAL AFFAIRS 2(2): 280-293. Fall 1972.

1432. Swidler, Joseph C. "The Challenge to State Regulation Agencies: The Experience of New York State." ANNALS OF THE AMERICAN ACADEMY OF POLITICAL AND SOCIAL SCIENCE 410( ): 106-119. November 1973.

1433. Szczelkan, Stefan A. SURVIVAL SCRAPBOOK 3: ENERGY. Schocken Books, New York. 1973.

1434. Szego, G.C. THE U.S. ENERGY PROBLEM. National Technical Information Service, Springfield, Virginia. 1971.

1435. Szyliowicz, Joseph and O'Neill, Bard (eds.). THE ENERGY CRISIS AND U.S. FOREIGN POLICY. Praeger, New York. (Forthcoming).

1436. Talbot, Albert R. POWER ALONG THE HUDSON: THE STORM KING CASE AND THE BIRTH OF ENVIRONMENTALISM. Dutton, New York. 1972.

1437. Tamplin, Arthur R. "Reacting to Reactors." TRIAL 10(1): 15-17. Janurary-February 1974.

1438. Tansil, John and Moyers, John C. "Residential Demand for Electricity." In ENERGY: DEMAND, CONSERVATION AND INSTITUTIONAL PROBLEMS, Michael Macrakis (ed.). M.I.T. Press, Cambridge, Massachusetts. 1974.

1439. Tarlock, A. Dan, Tippy, Roger and Francis, Frances Enseki. "Environmental Regulation of Power Plant Siting: Existing and Proposed Institutions." SOUTHERN CALIFORNIA LAW REVIEW 45: 502-569. Spring 1972.

1440. Taylor, Stewart F. "The Rapid Tramway: A Feasible Solution to the Urban Transportation Problem." CURRENT MUNICIPAL PROBLEMS 14(2): 131-148. Fall 1972.

1441. Taylor, Theodore B. and Willrich, Mason. NUCLEAR THEFT: RISKS AND SAFEGUARDS. Ballinger, Cambridge, Massachusetts. 1974.

1442. Teague, Elizabeth F. "Conference Report: Fuel Conservation and the Economy." ENERGY POLICY 2(2): 167- . June 1974.

1443. TECHNOLOGY REVIEW. "Energy Self-Sufficiency: An Economic Evaluation. Special Report of MIT Energy Laboratory." TECHNOLOGY REVIEW 76(6): 23- . May 1974.

1444. Teilhard de Chardin, Pierre. ACTIVATION OF ENERGY. Harcourt Brace Jovanovich, New York. 1971.

1445. Tell, William K., Jr. "Marine Sanctuaries: Balancing of Energy Vs. Environmental Needs." NATURAL RESOURCES LAWYER 6(1): 108-116. Winter 1973.

1446. Tennessee Valley Authority. TVA POWER, 1970. Tennessee Valley Authority, Knoxville, Kentucky. 1970.

1447. Tewksbury, J.G., et al. AN ANALYSIS OF THE REGULATORY ASPECTS OF FUEL OIL SUPPLY. Foster Associates, Washington, D.C. June 1973.

1448. Texas Eastern Transmission Corporation. ENERGY AND FUELS IN THE UNITED STATES, 1947-1980. Texas Eastern Transmission Corporation, Houston. 1961.

1449. Texas Eastern Transmission Corporation. COMPETITION AND GROWTH IN AMERICAN ENERGY MARKETS, 1947-1985. Texas Eastern Transmission Corporation, Houston. 1968.

1450. Thies, Austin C. "Water Quality and Power Plants." EDISON ELECTRIC INSTITUTE BULLETIN 37: 201-206. June-July 1969.

1451. Thirring, Hans. ENERGY FOR MAN: WINDMILLS TO NUCLEAR POWER. University of Indiana Press, Bloomington, Indiana. 1958.

1452. Thomas, John A.G. "Conference Report: Fuel and the Environment, London." ENERGY POLICY 2(1): 73. March 1974.

1453. Thomas, Trevor M. "World Energy Resources: Survey and Review." THE GEOGRAPHICAL REVIEW 63(2): 246-258. April 1973.

1454. Thompson, A.R. "A View From the North." CASE WESTERN RESERVE JOURNAL OF INTERNATIONAL LAW 5(1): 52-64. Winter 1972.

1455. Thompson, Dennis L. (ed.). POLITICS, POLICY AND NATURAL RESOURCES. Macmillan, New York. 1972.

1456. Thompson, Theos J. "Power Needs and Power Problems--Past, Present and Future." AMERICAN JOURNAL OF PUBLIC HEALTH 61: 1406-1415. July 1971.

1457. Thring, M.W. "Why the World Urgently Needs an Energy Policy." INTERNATIONAL AFFAIRS 4: 225-239. May 1973.

1458. Tihansky, Dennis P. COST ANALYSIS OF WATER POLLUTION CONTROL: AN ANNOTATED BIBLIOGRAPHY. U.S. Government Printing Office, Washington, D.C. 1973.

1459. Tihansky, Dennis P. "An Economic Assessment of Marine Water Pollution Damage." Paper Presented at the International Association for Pollution Control Conference, Pollution Control in Marine Industries. Montreal. June 1973.

1460. Tinic, S.M., Harnden, B.M. and Janssen, C.T.L. "Estimation of Rural Demand for Natural Gas." MANAGEMENT SCIENCE 20(4): 604-616. December 1973.

1461. Tiraspolsky, W. "Energy as the Key to Social Evolution." IMPACT OF SCIENCE ON SOCIETY 3(1): 5-17. 1952.

1462. Train, Russell E. "Energy Problems and Environmental Concern." SCIENCE AND PUBLIC AFFAIRS, BULLETIN OF THE ATOMIC SCIENTISTS 29(9): 43-47. November 1973.

1463. Train, Russell E. "The Long-Term Value of the Energy Crisis." THE FUTURIST 8(1): 14-18. February 1974.

1464. Treby, Elliott. "Jurisdictional Conflicts Over Offshore Oil and Gas." SCHOOL OF ADVANCED INTERNATIONAL STUDIES REVIEW 17(1): 20-24. Fall 1972.

1465. Trezise, Philip H. "The Outlook for Energy Supplies." DEPARTMENT OF STATE BULLETIN 63: 479-483. October 26, 1970.

1466. Trowbridge, G.F. "Environmental Issues in Reactor Licensing." ATOMIC ENERGY LAW JOURNAL 12: 251- . 1970.

1467. Tsivoglu, Ernest C. "Nuclear Power: The Social Conflict." ENVIRONMENTAL SCIENCE AND TECHNOLOGY 5: 404-410. May 1971.

1468. Tugwell, Franklin. "Petroleum Policy in Venezuela: Lessons in the Politics of Dependence Management." STUDIES IN COMPARATIVE INTERNATIONAL DEVELOPMENT 9(1): 84-120. Spring 1974.

1469. Turner, S.E., McCullough, C.R. and Lyerly, R.L. "Industrial Sabotage in Nuclear Power Plants." NUCLEAR SAFETY 11: 107-114. March-April 1970.

1470. Tussing, Arlon R. "Energy Resources and Demand." THE CONFERENCE BOARD RECORD 11(4): 18-23. April 1974.

1471. Twomey, James P. and Kuh, Peter G. "Governmental Programs, Resources and Regulatory Powers Available to Assist Localities During Coal Development." Region VIII, U.S. Department of Housing and Urban Development, Denver, Colorado. April 1974.

1472. Udall, Stewart L. "The Last Traffic Jam: Too Many Cars, Too Little Oil." ATLANTIC MONTHLY 230: 72-74. October 1972.

1473. Udall, Stewart L. "The Energy Crisis: A Radical Solution." WORLD 2: 34-36. May 8, 1973.

1474. Udell, Gilman G. FEDERAL POWER COMMISSION LAWS AND HYDROELECTRIC POWER DEVELOPMENT LAWS. U.S. Government Printing Office, Washington, D.C. 1971.

1475. Underwood, Joanna (ed.). THE PRICE OF POWER: ELECTRIC UTILITIES AND THE ENVIRONMENT. Council on Economic Priorities, New York. 1972.

1476. United Nations. "World Energy Requirements in 1975 and 2000." In PROCEEDINGS OF THE INTERNATIONAL CONFERENCE ON THE PEACEFUL USES OF ATOMIC ENERGY, VOLUME I, Held in Geneva, August 1955, United Nations, New York. 1956. (pp. 3-33).

1477. United Nations. Department of Economic and Social Affairs. WORLD ENERGY SUPPLIES, 1964-1968. United Nations, New York. 1970.

1478. United Nations. Department of Economic and Social Affairs. WORLD ENERGY SUPPLIES, 1961-1970. United Nations, New York. 1972.

1479. United Nations. Economic and Social Council. NEW SOURCES OF ENERGY AND ENERGY DEVELOPMENT: SOLAR ENERGY, WIND POWER, GEOTHERMAL ENERGY. United Nations, New York. 1972.

1480. UNESCO COURIER. "The Big Five of World Energy: Coal, Petroleum, Natural Gas, Uranium, Hydro-Power." UNESCO COURIER (1): 6-13. January 1974.

1481. U.S. Atomic Energy Commission. POTENTIAL NUCLEAR POWER GROWTH PATTERNS. U.S. Atomic Energy Commission, Oak Ridge, Tennessee. 1970.

1482. U.S. Atomic Energy Commission. FORECAST OF GROWTH OF NUCLEAR POWER. U.S. Atomic Energy Commission, Oak Ridge, Tennessee. 1971.

1483. U.S. Atomic Energy Commission. NUCLEAR POWER 1973-2000. U.S. Government Printing Office, Washington, D.C. 1972.

1484. U.S. Atomic Energy Commission. UPDATED COST-BENEFIT ANALYSIS: U.S. BREEDER REACTOR PROGRAM. U.S. Government Printing Office, Washington, D.C. 1972.

1485. U.S. Atomic Energy Commission. THE NUCLEAR INDUSTRY 1973. U.S. Government Printing Office, Washington, D.C. 1973.

1486. U.S. Atomic Energy Commission. NUCLEAR REACTORS BUILT, BEING BUILT OR PLANNED IN THE U.S. AS OF JUNE 30, 1973. National Technical Information Service, Springfield, Virginia. 1973.

1487. U.S. Atomic Energy Commission. Division of Controlled Thermonuclear Research. FUSION POWER, RESEARCH AND DEVELOPMENT REQUIREMENTS. U.S. Government Printing Office, Washington, D.C. 1973.

1488. U.S. Atomic Energy Commission. Division of Industrial Participation. THE NUCLEAR INDUSTRY, 1969. U.S. Government Printing Office, Washington, D.C. 1969.

1489. U.S. Atomic Energy Commission. Division of Reactor Development. COST-BENEFIT ANALYSIS OF THE U.S. BREEDER REACTOR PROGRAM. U.S. Government Printing Office, Washington, D.C. 1969.

1490. U.S. Atomic Energy Commission. Divison of Reactor Development and Technology. THERMAL EFFECTS AND U.S. NUCLEAR POWER STATIONS. U.S. Government Printing Office, Washington, D.C. 1971.

1491. U.S. Atomic Energy Commission. Divison of Technical Information. ABUNDANT NUCLEAR ENERGY. AEC Symposium Series Number 14. U.S. Atomic Energy Commission, Oak Ridge, Tennessee. 1969.

1492. U.S. Bureau of Indian Affairs in cooperation with the Tribes of the Northern Plains. "Indians in the Northern Great Plains; Anticipated Socio-Economic Impacts of Coal Development." U.S. Bureau of Indian Affairs, Billings, Montana. April 1974.

1493. U.S. Bureau of Mines. ENERGY PRODUCTION AND CONSUMPTION IN THE UNITED STATES: AN ANALYTICAL STUDY BASED ON 1954 DATA. U.S. Government Printing Office, Washington, D.C. 1961.

1494. U.S. Bureau of Mines. TRANSPORTATION COSTS OF FOSSIL FUELS. U.S. Government Printing Office, Washington, D.C. 1971.

1495. U.S. Bureau of Mines. COST ANALYSES OF MODEL MINES FOR STRIP MINING OF COAL IN THE UNITED STATES. Information Circular Number 8535. U.S. Government Printing Office, Washington, D.C. 1972.

1496. U.S. Bureau of Mines. U.S. ENERGY THROUGH THE YEAR 2000. U.S. Government Printing Office, Washington, D.C. 1972.

1497. U.S. Bureau of Natural Gas. NATIONAL GAS SUPPLY AND DEMAND 1971-1990. U.S. Government Printing Office, Washington, D.C. 1972.

1498. U.S. Bureau of Reclamation and Montana State University Center for Interdisciplinary Studies. "The Anticipated Effects of Major Coal Development on Public Services, Costs and Revenues in Six Selected Counties." Center for Interdisciplinary Studies, Montana State University, Missoula and U.S. Bureau of Reclamation, Billings, Montana. April 1974.

1499. U.S. Cabinet Task Force on Oil Import Controls. THE OIL IMPORT QUESTION: A REPORT ON THE RELATIONSHIP OF OIL IMPORTS TO THE NATIONAL SECURITY. U.S. Government Printing Office, Washington, D.C. 1970.

1500. U.S. Citizens Advisory Commission on Environmental Quality. CITIZEN ACTION GUIDE TO ENERGY CONSERVATION. U.S. Government Printing Office, Washington, D.C. 1973.

1501. U.S. Commission on the Organization of the Executive Branch of the Government. WATER RESOURCES AND POWER. U.S. Government Printing Office, Washington, D.C. 1955.

1502. U.S. Congress. DEVELOPMENT GROWTH AND STATE OF ATOMIC ENERGY INDUSTRY. Hearings before the Joint Committee on Atomic Energy, August 10-September 8, 1965. U.S. Government Printing Office, Washington, D.C. 1965.

1503. U.S. Congress. PARTICIPATION BY SMALL ELECTRIC UTILITIES IN NUCLEAR POWER. Hearings before the Joint Committee on Atomic Energy, April 30-May 3, 1968; June 11-13, 1968. U.S. Government Printing Office, Washington, D.C. 1968.

1504. U.S. Congress. A REVIEW OF ENERGY ISSUES AND THE 91st CONGRESS. U.S. Government Printing Office, Washington, D.C. 1970.

1505. U.S. Congress. TRENDS AND GROWTH PROJECTIONS OF THE ELECTRIC POWER INDUSTRY, ENVIRONMENTAL EFFECTS OF PRODUCING ELECTRIC POWER. Hearings before the Joint Committee on Atomic Energy. 91st Congress. U.S. Government Printing Office, Washington, D.C. 1970.

1506. U.S. Congress. EMERGENCY PETROLEUM ALLOCATION ACT. S. 1570. Approved November 27, 1973. U.S. Government Printing Office, Washington, D.C. 1973.

1507. U.S. Congress. Conference Committees. NATIONAL ENVIRONMENTAL POLICY ACT OF 1969. Report to accompany S. 1075. 91st Congress. U.S. Government Printing Office, Washington, D.C. 1969.

1508. U.S. Congress. Conference Committees. RESOURCE RECOVERY ACT OF 1970. Report to accompany H.R. 11833. 91st Congress. U.S. Government Printing Office, Washington, D.C. 1970.

1509. U.S. Congress. House. MANDATORY OIL IMPORT CONTROL PROGRAMS, ITS IMPACT UPON DOMESTIC MINERALS INDUSTRY AND NATIONAL SECURITY. Hearings before the Subcommittee on Mines and Mining of the Committee on Interior and Insular Affairs. 90th Congress. U.S. Government Printing Office, Washington, D.C. 1968.

1510. U.S. Congress. House. EFFECTS OF POPULATION GROWTH ON NATURAL RESOURCES AND THE ENVIRONMENT. Hearings before the Subcommittee on Conservation and Natural Resources of the Committee on Government Operations. 91st Congress. U.S. Government Printing Office, Washington, D.C. 1969.

1511. U.S. Congress. House. THE ENVIRONMENTAL DECADE: ACTION PROPOSALS FOR THE 1970's. Report of the Subcommittee on Conservation and Natural Resources of the Committee on Government Operations. 91st Congress. U.S. Government Printing Office, Washington, D.C. 1970.

1512. U.S. Congress. House. ESTABLISHING A NATIONAL MINING AND MINERALS POLICY. Report of the Committee on Interior and Insular Affairs to accompany S. 719. 91st Congress. U.S. Government Printing Office, Washington, D.C. 1970.

1513. U.S. Congress. House. THE IMPACT OF THE ENERGY AND FUEL CRISIS ON SMALL BUSINESS. Hearings before the Subcommittee on Special Small Business Problems of the Select Committee on Small Business. 91st Congress. U.S. Government Printing Office, Washington, D.C. 1970.

1514. U.S. Congress. House. OIL IMPORT CONTROLS. Hearings before the Subcommittee on Mines and Mining of the Committee on Interior and Insular Affairs. 91st Congress. U.S. Government Printing Office, Washington, D.C. 1970.

1515. U.S. Congress. House. REPORT ON THE OIL IMPORT QUESTION: TOGETHER WITH DISSENTING AND SEPARATE VIEWS. Report of the Subcommittee on Mines and Mining of the Committee on Interior and Insular Affairs. 91st Congress. U.S. Government Printing Office, Washington, D.C. 1970.

1516. U.S. Congress. House. ENERGY RESEARCH AND DEVELOPMENT. Hearings before the Subcommittee on Science, Research and Development of the Committee on Science and Astronautics. 92nd Congress. U.S. Government Prining Office, Washington, D.C. 1972.

1517. U.S. Congress. House. CONSERVATION AND EFFICIENT USE OF ENERGY. Joint hearings before certain Subcommittees of the Committee on Government Operations and Science and Astronautics. 93rd Congress. U.S. Government Printing Office, Washington, D.C. 1973.

1518. U.S. Congress. House. ENERGY REORGANIZATION ACT OF 1973. Hearings before a Subcommittee of the Committee on Government Operations. 93rd Congress. U.S. Government Printing Office, Washington, D.C. 1973.

1519. U.S. Congress. House. ENERGY RESEARCH AND DEVELOPMENT. OVERVIEW OF THE NATIONAL EFFORT. Hearings before the Subcommittee on Energy of the Committee on Science and Astronautics. 93rd Congress. U.S. Government Printing Office, Washington, D.C. 1973.

1520. U.S. Congress. House. SHORT TERM ENERGY SHORTAGES. Hearings before the Committee on Science and Astronautics. 93rd Congress. U.S. Government Printing Office, Washington, D.C. 1973.

1521. U.S. Congress. House. SMALL BUSINESS AND THE ENERGY SHORTAGE. Hearings before the Subcommittee on Special Small Business Problems of the Select Committee on Small Business. 93rd Congress. U.S. Government Printing Office, Washington, D.C. 1973.

1522. U.S. Congress. House. AGRICULTURE AND THE FUEL CRISIS. Hearings before the Committee on Agriculture. 93rd Congress. U.S. Government Printing Office, Washington, D.C. 1974.

1523. U.S. Congress. House. FEDERAL ENERGY ADMINISTRATION. Hearings before a Subcommittee on the Committee on Government Operations. 93rd Congress. U.S. Government Printing Office, Washington, D.C. 1974.

1524. U.S. Congress. House. ENERGY EMERGENCY ACT. Report to accompany S. 2584, February 7, 1974. 93rd Congress. U.S. Government Printing Office, Washington, D.C. 1974.

1525. U.S. Congress. House. Committee on Interior and Insular Affairs. ENERGY "DEMAND" STUDIES: AN ANALYSIS AND APPRAISAL. U.S. Government Printing Office, Washington, D.C. 1972.

1526. U.S. Congress. House. Committee on Interior and Insular Affairs. SELECTED READINGS ON THE FUELS AND ENERGY CRISIS. U.S. Government Printing Office, Washington, D.C. 1972.

1527. U.S. Congress. House. Committee on Interior and Insular Affairs. Subcommittee on the Environment. AMERICA'S ENERGY POTENTIAL: A SUMMARY AND EXPLANATION. U.S. Government Printing Office, Washington, D.C. 1973.

1528. U.S. Congress. House. Committee on Interstate and Foreign Commerce. Subcommittee on Communications and Power. BILLS RELATING TO POWERPLANT SITING AND ENVIRONMENTAL PROTECTION. U.S. Government Printing Office, Washington, D.C. 1971.

1529. U.S. Congress. House. Committee on Science and Astronautics. NUCLEAR POWER SOURCES. Staff study for Subcommittee on NASA Oversight. U.S. Government Printing Office, Washington, D.C. 1967.

1530. U.S. Congress. House. Committee on Science and Astronautics. Task Force on Energy. ENERGY--THE ULTIMATE RESOURCE. U.S. Government Printing Office, Washington, D.C. 1971.

1531. U.S. Congress. Joint Committee on Atomic Energy. SELECTED MATERIALS ON ENVIRONMENTAL EFFECTS OF PRODUCING ELECTRIC POWER. U.S. Government Printing Office, Washington, D.C. 1969.

1532. U.S. Congress. Joint Committee on Atomic Energy. NUCLEAR POWER AND RELATED ENERGY PROBLEMS, 1968 THROUGH 1970. U.S. Government Printing Office, Washington, D.C. 1971.

1533. U.S. Congress. Joint Committee on Atomic Energy. CERTAIN BACKGROUND INFORMATION FOR CONSIDERATION WHEN EVALUATING THE "NATIONAL ENERGY DILEMMA." U.S. Government Printing Office, Washington, D.C. 1973.

1534. U.S. Congress. Joint Committee on Atomic Energy. UNDERSTANDING THE NATIONAL ENERGY DILEMMA. Staff Report. 93rd Congress. U.S. Government Printing Office, Washington, D.C. 1973.

1535. U.S. Congress. Joint Economic Committee. THE ECONOMY, ENERGY AND THE ENVIRONMENT. U.S. Government Printing Office, Washington, D.C. September 1970.

1536. U.S. Congress. Joint Economic Committee. CONSERVATION OF ENERGY. U.S. Government Printing Office, Washington, D.C. 1972.

1537. U.S. Congress. Joint Economic Committee. Subcommittee on International Economics. ECONOMIC IMPACT OF THE PETROLEUM SHORTAGE. U.S. Government Printing Office, Washington, D.C. 1974.

1538. U.S. Congress. Senate. ELECTRIC POWER RELIABILITY. Hearings before the Committee on Commerce. 90th Congress. U.S. Government Printing Office, Washington, D.C. 1967.

1539. U.S. Congress. Senate. ELECTRIC VEHICLES AND OTHER ALTERNATIVES TO INTERNAL COMBUSTION ENGINES. Joint Hearings before the Subcommittee on Air and Water Pollution of the Committee on Public Works and the Committee on Commerce. 90th Congress. U.S. Government Printing Office, Washington, D.C. 1967.

1540. U.S. Congress. Senage. MINERAL SHORTAGES. Hearings before the Subcommittee on Minerals, and Fuels of the Committee on Interior and Insular Affairs. 90th Congress. U.S. Government Printing Office, Washington, D.C. 1968.

1541. U.S. Congress. Senate. ESTABLISHING A NATIONAL MINERALS POLICY. Report of the Committee on Interior and Insular Affairs to accompany S. 719. 91st Congress. U.S. Government Printing Office, Washington, D.C. 1969.

1542. U.S. Congress. Senate. GOVERNMENTAL INTERVENTION IN THE MARKET MECHANISM: THE PETROLEUM INDUSTRY, PART I. Hearings before the Subcommittee on Antitrust and Monopoly of the Committee on the Judiciary. 91st Congress. U.S. Government Printing Office, Washington, D.C. 1969.

1543. U.S. Congress. Senate. GOVERNMENTAL INTERVENTION IN THE MARKET MECHANISM: THE PETROLEUM INDUSTRY, PART II. Hearings before the Subcommittee on Antitrust and Monopoly of the Committee on the Judiciary. 91st Congress. U.S. Government Printing Office, Washington, D.C. 1969.

1544. U.S. Congress. Senate. GOVERNMENTAL INTERVENTION IN THE MARKET MECHANISM: THE PETROLEUM INDUSTRY, PART III. Hearings before the Subcommittee on Antitrust and Monopoly of the Committee on the Judiciary. 91st Congress. U.S. Governming Printing Office, Washington, D.C. 1969.

1545. U.S. Congress. Senate. NATIONAL ENVIRONMENTAL POLICY ACT OF 1969. Report of the Committee on Interior and Insular Affairs to accompany S. 1075. 91st Congress. U.S. Government Printing Office, Washington, D.C. 1969.

1546. U.S. Congress. Senate. TOWARD A NATIONAL MATERIALS POLICY. Report of the Committee on Public Works on a Proposed Commission on National Materials Policy. 91st Congress. U.S. Government Printing Office, Washington, D.C. 1969.

1547. U.S. Congress. Senate. A BILL TO ESTABLISH A COMMISSION OF FUELS AND ENERGY. S. 4092. Hearings before the Committee on Interior and Insular Affairs. 91st Congress. U.S. Government Printing Office, Washington, D.C. 1970.

1548. U.S. Congress. Senate. ENVIRONMENTAL PROTECTION ACT OF 1970. Hearings before the Subcommittee on Energy, Natural Resources and the Environment of the Committee on Commerce. 91st Congress. U.S. Government Printing Office, Washington, D.C. 1970.

1549. U.S. Congress. Senate. FEDERAL POWER COMMISSION OVERSIGHT. Hearings before the Subcommittee on Energy, Natural Resources and the Environment of the Committee on Commerce. 91st Congress. U.S. Government Printing Office, Washington, D.C. 1970.

1550. U.S. Congress. Senate. FUELS AND ENERGY. Hearings before the Subcommittee on Minerals, Materials, and Fuels of the Committee on Interior and Insular Affairs. 91st Congress. U.S. Government Printing Office, Washington, D.C. 1970.

1551. U.S. Congress. Senate. NATURAL GAS SUPPLY STUDY. Hearings before the Subcommittee on Minerals, Materials, and Fuels of the Committee on Interior and Insular Affairs. 91st Congress. U.S. Government Printing Office, Washington, D.C. 1970.

1552. U.S. Congress. Senate. THE OIL SITUATION IN THE NORTHEAST AND GREAT LAKES REGION. Hearings before the Subcommittee on Small Business of the Committee on Banking and Currency. 91st Congress. U.S. Government Printing Office, Washington, D.C. 1970.

1553. U.S. Congress. Senate. SOME ENVIRONMENTAL IMPLICATIONS OF NATIONAL FUELS POLICIES. Report of the Committee on Public Works. 91st Congress. U.S. Government Printing Office, Washington, D.C. 1970.

1554. U.S. Congress. Senate. COMPETITIVE ASPECTS OF THE ENERGY INDUSTRY. Hearings before the Subcommittee on Antitrust and Monopoly of the Committee on the Judiciary. 91st Congress. U.S. Government Printing Office, Washington, D.C. 1971.

1555. U.S. Congress. Senate. NATIONAL FUELS AND ENERGY POLICY. Hearings before the Committee on Interior and Insular Affairs. 92nd Congress. U.S. Government Printing Office, Washington, D.C. 1971.

1556. U.S. Congress. Senate. THE PRESIDENT'S ENERGY MESSAGE. Hearings before the Committee on Interior and Insular Affairs. 92nd Congress. U.S. Government Printing Office, Washington, D.C. 1971.

1557. U.S. Congress. Senate. ADVANCED POWER CYCLES. Hearings before the Committee on Interior and Insular Affairs. 92nd Congress. U.S. Government Printing Office, Washington, D.C. 1972.

1558. U.S. Congress. Senate. DEEP WATER PORT POLICY ISSUES. Hearings before the Committee on Interior and Insular Affairs. 92nd Congress. U.S. Government Printing Office, Washington, D.C. 1972.

1559. U.S. Congress. Senate. ENERGY RESEARCH POLICY ALTERNATIVES. Hearings before the Committee on Interior and Insular Affairs. 92nd Congress. U.S. Government Printing Office, Washington, D.C. 1972.

1560. U.S. Congress. Senate. FEDERAL LEASING AND DISPOSAL POLICIES. Hearings before the Committee on Interior and Insular Affairs. 92nd Congress. U.S. Government Printing Office, Washington, D.C. 1972.

-111-

1561. U.S. Congress. Senate. NATURAL GAS POLICY ISSUES. Hearings before the Committee on Interior and Insular Affairs. 92nd Congress. U.S. Government Printing Office, Washington, D.C. 1972.

1562. U.S. Congress. Senate. OUTER CONTINENTAL SHELF POLICY ISSUES. Hearings before the Committee on Interior and Insular Affairs. 92nd. Congress. U.S. Government Printing Office, Washington, D.C. 1972.

1563. U.S. Congress. Senate. PROBLEMS OF ELECTRICAL POWER PRODUCTION IN THE SOUTHWEST. Hearings before the Committee on Interior and Insular Affairs. 92nd Congress. U.S. Government Printing Office, Washington, D.C. 1972.

1564. U.S. Congress. Senate. SURFACE MINING. Hearings before the Committee on Interior and Insular Affairs. 92nd Congress. U.S. Government Printing Office, Washington, D.C. 1972.

1565. U.S. Congress. Senate. TO ESTABLISH A DEPARTMENT OF NATURAL RESOURCES. L. 2410. Hearings before the Committee on Interior and Insular Affairs. 92nd Congress. U.S. Government Printing Office, Washington, D.C. 1972.

1566. U.S. Congress. Senate. TRENDS IN OIL AND GAS EXPLORATION. Hearings before the Committee on Interior and Insular Affairs. 92nd Congress. U.S. Government Printing Office, Washington, D.C. 1972.

1567. U.S. Congress. Senate. COAL POLICY ISSUES. Hearings before the Committee on Interior and Insular Affairs. 93rd Congress. U.S. Government Printing Office, Washington D.C. 1973.

1568. U.S. Congress. Senate. COMPETITION IN THE ENERGY INDUSTRY. Hearings before the Committee on the Judiciary. 93rd Congress. U.S. Government Printing Office, Washington, D.C. 1973.

1569. U.S. Congress. Senate. CONSERVATION OF ENERGY. Hearings before the Committee on Interior and Insular Affairs. 93rd Congress. U.S. Government Printing Office, Washington, D.C. 1973.

1570. U.S. Congress. Senate. CURRENT FUEL SHORTAGES. Hearings before the Committee on Interior and Insular Affaris. 93rd Congress. U.S. Government Printing Office, Washington, D.C. 1973.

1571. U.S. Congress. Senate. ENERGY EMERGENCY LEGISLATION PART I. Hearings before the Committee on Interior and Insular Affairs. 93rd Congress. U.S. Government Printing Office, Washington, D.C. 1973.

1572. U.S. Congress. Senate. FINANCIAL REQUIREMENTS OF THE NATION'S ENERGY INDUSTRIES. Hearings before the Committee on Interior and Insular Affairs. 93rd Congress. U.S. Government Printing Office, Washington, D.C. 1973.

1573. U.S. Congress. Senate. FISCAL POLICY AND THE ENERGY CRISIS. Briefing material prepared for the Subcommittee on Energy of the Finance Committee. 93rd Congress. U.S. Government Printing Office, Washington, D.C. 1973.

1574. U.S. Congress. Senate. FUEL SHORTAGES. Hearings before the Committee on Interior and Insular Affairs. 93rd Congress. U.S. Government Printing Office, Washington, D.C. 1973.

1575. U.S. Congress. Senate. THE NATIONAL RESEARCH AND DEVELOPMENT POLICY ACT OF 1973. Hearings before the Committee on Interior and Insular Affairs. 93rd Congress. U.S. Government Printing Office, Washington, D.C. 1973.

1576. U.S. Congress. Senate. OIL AND GAS IMPORT ISSUES. Hearings before the Committee on Interior and Insular Affairs. 93rd Congress. U.S. Government Printing Office, Washington, D.C. 1973.

1577. U.S. Congress. Senate. OVERSIGHT AND EFFICIENCY OF EXECUTIVE AGENCIES WITH RESPECT TO THE PETROLEUM INDUSTRY, ESPECIALLY AS IT RELATES TO FUEL SHORTAGES. Staff Study of the Permanent Subcommittee on Investigations of the Committee on Government Operations. 93rd Congress. U.S. Government Printing Office, Washington, D.C. 1973.

1578. U.S. Congress. Senate. PRESIDENT'S ENERGY MESSAGE OF 1973 AND S. 1570, THE EMERGENCY FUEL AND ALLOCATION ACT OF 1973. Hearings before the Committee on Interior and Insular Affairs. 93rd Congress. U.S. Government Printing Office, Washington, D.C. 1973.

1579. U.S. Congress. Senate. STRATEGIC PETROLEUM RESERVES, S. 1586. Hearings before the Committee on Interior and Insular Affairs. 93rd Congress. U.S. Government Printing Office, Washington, D.C. 1973.

1580. U.S. Congress. Senate. ENERGY AND FOREIGN POLICY. Hearings before the Committee on Foreign Relations. 93rd Congress. U.S. Government Printing Office, Washington, D.C. 1974.

1581. U.S. Congress. Senate. ENERGY INFORMATION ACT. Hearings before the Committee on Interior and Insular Affairs. 93rd Congress. U.S. Government Printing Office, Washington, D.C. 1974.

-113-

1582. U.S. Congress. Senate. HEARINGS TO ESTABLISH A DEPARTMENT OF ENERGY AND NATURAL RESOURCES. Hearings before the Committee on Government Operations. 93rd Congress. U.S. Government Printing Office, Washington, D.C. 1974.

1583. U.S. Congress. Senate. MARKET PERFORMANCE AND COMPETITION IN THE PETROLEUM INDUSTRY. Hearings before the Committee on Interior and Insular Affairs. 93rd Congress. U.S. Government Printing Office, Washington, D.C. 1974.

1584. U.S. Congress. Senate. THE NATIONAL COAL CONVERSION ACT. Hearings before the Committee on Interior and Insular Affairs. 93rd Congress. U.S. Government Printing Office, Washington, D.C. 1974.

1585. U.S. Congress. Senate. OVERSIGHT--MANDATORY PETROLEUM ALLOCATION PROGRAMS. Hearings before the Committee on Interior and Insular Affairs. 93rd Congress. U.S. Government Printing Office, Washington, D.C. 1974.

1586. U.S. Congress. Senate. Committee on Government Operations. Sub-Committee on Intergovernmental Relations. INTERGOVERNMENTAL COORDINATION OF POWER DEVELOPMENT AND ENVIRONMENTAL PROTECTION ACT. U.S. Government Printing Office, Washington, D.C. 1970.

1587. U.S. Congress. Senate. Committee on Interior and Insular Affairs. ESTABLISHING A NATIONAL MINERALS POLICY. 91st Congress. U.S. Government Printing Office, Washington, D.C. 1969.

1588. U.S. Congress. Senate. Committee on Interior and Insular Affairs. A BIBLIOGRAPHY OF CONGRESSIONAL PUBLICATIONS ON ENERGY FROM THE 89th CONGRESS TO JULY 1, 1971. 92nd Congress. U.S. Government Printing Office, Washington, D.C. 1971.

1589. U.S. Congress. Senate. Committee on Interior and Insular Affairs. A BIBLIOGRAPHY OF NONTECHNICAL LITERATURE ON ENERGY. 92nd Congress. U.S. Government Printing Office, Washington, D.C. 1971.

1590. U.S. Congress. Senate. Committee on Interior and Insular Affairs. CONSIDERATIONS IN THE FORMULATION OF NATIONAL ENERGY POLICY. 92nd Congress. U.S. Government Printing Office, Washington, D.C. 1971.

1591. U.S. Congress. Senate. Committee on Interior and Insular Affairs. ENERGY POLICY AND NATIONAL GOALS: A SYMPOSIUM. 92nd Congress. U.S. Government Printing Office, Washington, D.C. 1971.

1592. U.S. Congress. Senate. Committee on Interior and Insular Affairs. THE EVOLUTION AND DYNAMICS OF NATIONAL GOALS IN THE UNITED STATES. 92nd Congress. U.S. Government Printing Office, Washington, D.C. 1971.

1593. U.S. Congress. Senate. Committee on Interior and Insular Affairs. GOALS AND OBJECTIVES OF FEDERAL AGENCIES IN FUELS AND ENERGY. 92nd Congress. U.S. Government Printing Office, Washington, D.C. 1971.

1594. U.S. Congress. Senate. Committee on Interior and Insular Affairs. THE ISSUES RELATED TO SURFACE MINING: A SUMMARY REVIEW, WITH SELECTED READINGS. 92nd Congress. U.S. Government Printing Office, Washington, D.C. 1971.

1595. U.S. Congress. Senate. Committee on Interior and Insular Affairs. LEGISLATIVE HISTORY OF S. RES. 45: A NATIONAL FUELS AND ENERGY POLICY STUDY. 92nd Congress. U.S. Government Printing Office, Washington, D.C. 1971.

1596. U.S. Congress. Senate. Committee on Interior and Insular Affairs. REVIEW OF ENERGY ISSUES: 92nd CONGRESS. U.S. Government Printing Office, Washington, D.C. 1971.

1597. U.S. Congress. Senate. Committee on Interior and Insular Affairs. A REVIEW OF THE ENERGY RESOURCES OF THE PUBLIC LANDS, BASED ON STUDIES SPONSORED BY THE PUBLIC LAND LAW REVIEW COMMISSION. 92nd Congress. U.S. Government Printing Office, Washington, D.C. 1971.

1598. U.S. Congress. Senate. Committee on Interior and Insular Affairs. SELECTED READINGS ON ECONOMIC GROWTH IN RELATION TO POPULATION INCREASE, NATURAL RESOURCE AVAILABILITY, ENVIRONMENTAL CONTROL AND ENERGY NEEDS. 92nd Congress. U.S. Government Printing Office, Washington, D.C. 1971.

1599. U.S. Congress. Senate. Committee on Interior and Insular Affairs. STUDIES AND REPORTS RELEVANT TO NATIONAL ENERGY POLICIES. 92nd Congress. U.S. Government Printing Office, Washington, D.C. 1971.

1600. U.S. Congress. Senate. Committee on Interior and Insular Affairs. EFFECTS OF CALVERT CLIFFS AND OTHER COURT DECISIONS UPON NUCLEAR POWER IN THE UNITED STATES. 92nd Congress. U.S. Government Printing Office, Washington, D.C. 1972.

1601. U.S. Congress. Senate. Committee on Interior and Insular Affairs. FEDERAL RESOURCES (FUNDING AND PERSONNEL) IN ENERGY RELATED ACTIVITIES, FISCAL YEARS 1972 AND 1973. 92nd Congress. U.S. Government Printing Office, Washington, D.C. 1972.

1602. U.S. Congress. Senate. Committee on Interior and Insular Affairs. PROBLEMS OF ELECTRICAL POWER PRODUCTION IN THE SOUTHWEST. 92nd Congress. U.S. Government Printing Office, Washington, D.C. 1972.

1603. U.S. Congress. Senate. Committee on Interior and Insular Affairs. SUMMARY REPORT OF THE CORNELL WORKSHOP ON ENERGY AND THE ENVIRONMENT. February 22-24. 92nd Congress. U.S. Government Printing Office, Washington, D.C. 1972.

1604. U.S. Congress. Senate. Committee on Interior and Insular Affairs. SUPPLEMENTAL BIBLIOGRAPHY OF PUBLICATIONS ON ENERGY. 92nd Congress. U.S. Government Printing Office, Washington, D.C. 1972.

1605. U.S. Congress. Senate. Committee on Interior and Insular Affairs. SURVEY OF ENERGY CONSUMPTION PROJECTIONS. 92nd Congress. U.S. Government Printing Office, Washington, D.C. 1972.

1606. U.S. Congress. Senate. Committee on Interior and Insular Affairs. COAL SURFACE MINING AND RECLAMATION. 93rd Congress. U.S. Government Printing Office, Washington, D.C. 1973.

1607. U.S. Congress. Senate. Committee on Interior and Insular Affairs. ENERGY RESEARCH AND DEVELOPMENT: PROBLEMS AND PROSPECTS. 93rd. Congress. U.S. Government Printing Office, Washington, D.C. 1973.

1608. U.S. Congress. Senate. Committee on Interior and Insular Affairs. FACTORS AFFECTING THE USE OF COAL IN PRESENT AND FUTURE ENERGY MARKETS. 93rd Congress. U.S. Government Printing Office, Washington, D.C. 1973.

1609. U.S. Congress. Senate. Committee on Interior and Insular Affairs. HISTORY OF FEDERAL ENERGY ORGANIZATIONS. 93rd Congress. U.S. Government Printing Office, Washington, D.C. 1973.

1610. U.S. Congress. Senate. Committee on Interior and Insular Affairs. THE GASOLINE SHORTAGE: A NATIONAL PERSPECTIVE. 93rd Congress. U.S. Government Printing Office, Washington, D.C. 1973.

1611. U.S. Congress. Senate. Committee on Interior and Insular Affairs. LEGISLATIVE AUTHORITY OF FEDERAL AGENCIES WITH RESPECT TO FUELS AND ENERGY. 93rd Congress. U.S. Government Printing Office, Washington, D.C. 1973.

1612. U.S. Congress. Senate. Committee on Interior and Insular Affairs. NATURAL GAS POLICY ISSUES AND OPTIONS. 93rd Congress. U.S. Government Printing Office, Washington, D.C. 1973.

1613. U.S. Congress. Senate. Committee on Interior and Insular Affairs. PRELIMINARY FEDERAL TRADE COMMISSION REPORT ON ITS INVESTIGATION OF THE PETROLEUM INDUSTRY. 93rd Congress. U.S. Government Printing Office, Washington, D.C. 1973.

-116-

1614. U.S. Congress. Senate. Committee on Interior and Insular Affairs.
PRESIDENTIAL ENERGY STATEMENTS. 93rd Congress. U.S. Government Printing Office, Washington, D.C. 1973.

1615. U.S. Congress. Senate. Committee on Interior and Insular Affairs.
REPORT: NATIONAL EMERGENCY ENERGY ACT OF 1973. 93rd Congress. U.S. Government Printing Office, Washington, D.C. 1973.

1616. U.S. Congress. Senate. Committee on Interior and Insular Affairs.
REPORT: NATIONAL FUELS AND ENERGY CONSERVATION ACT OF 1973. 93rd Congress. U.S. Government Printing Office, Washington, D.C. 1973.

1617. U.S. Congress. Senate. Committee on Interior and Insular Affairs.
A REVIEW OF ENERGY POLICY ACTIVITIES OF THE 92nd CONGRESS. 93rd Congress. U.S. Government Printing Office, Washington, D.C. 1973.

1618. U.S. Congress. Senate. Committee on Interior and Insular Affairs.
SUMMARY OF THE ENERGY CONSERVATION AND DEVELOPMENT RECOMMENDATIONS CONTAINED IN THE FINAL REPORT OF THE NATIONAL COMMISSION ON MATERIALS POLICY. 93rd Congress. U.S. Government Printing Office, Washington, D.C. 1973.

1619. U.S. Congress. Senate. Committee on Interior and Insular Affairs.
TOWARD A RATIONAL POLICY FOR OIL AND GAS IMPORTS: A POLICY BACKGROUND PAPER. 93rd Congress. U.S. Government Printing Office, Washington, D.C. 1973.

1620. U.S. Congress. Senate. Committee on Interior and Insular Affairs.
AN ANALYSIS OF THE FEDERAL TAX TREATMENT OF OIL AND GAS AND SOME POLICY ALTERNATIVES. 93rd Congress. U.S. Government Printing Office, Washington, D.C. 1974.

1621. U.S. Congress. Senate. Committee on Interior and Insular Affairs.
AN ASSESSMENT AND ANALYSIS OF THE ENERGY EMERGENCY. 93rd Congress. U.S. Government Printing Office, Washington, D.C. 1974.

1622. U.S. Congress. Senate. Committee on Interior and Insular Affairs.
CURRENT ANALYSES OF PETROLEUM SUPPLIES FOR FIRST QUARTER 1974. 93rd Congress. U.S. Government Printing Office, Washington, D.C. 1974.

1623. U.S. Congress. Senate. Committee on Interior and Insular Affairs.
DEEPWATER PORT POLICY ISSUES. 93rd Congress. U.S. Government Printing Office, Washington, D.C. 1974.

1624. U.S. Congress. Senate. Committee on Interior and Insular Affairs.
ESTIMATES AND ANALYSIS OF FUEL SUPPLY OUTLOOK FOR 1974. 93rd Congress. U.S. Government Printing Office, Washington, D.C. 1974.

-117-

1625. U.S. Congress. Senate. Committee on Interior and Insular Affairs. FEDERAL CHARTERS FOR ENERGY CORPORATIONS--SELECTED MATERIALS. 93rd Congress. U.S. Government Printing Office, Washington, D.C. 1974.

1626. U.S. Congress. Senate. Committee on Interior and Insular Affairs. HIGHLIGHTS OF ENERGY LEGISLATION IN THE 93rd CONGRESS, 1ST SESSION. 93rd Congress. U.S. Government Printing Office, Washington, D.C. 1974.

1627. U.S. Congress. Senate. Committee on Interior and Insular Affairs. IMPLICATIONS OF RECENT OPEC OIL PRICE INCREASES. 93rd Congress. U.S. Government Printing Office, Washington, D.C. 1974.

1628. U.S. Congress. Senate. Committee on Interior and Insular Affairs. PETROLEUM ALLOCATION POLICIES AND PROGRAMS. 93rd Congress. U.S. Government Printing Office, Washington, D.C. 1974.

1629. U.S. Congress. Senate. Committee on Interior and Insular Affairs. THE PROSPECTS FOR GASOLINE AVAILABILITY: 1974. 93rd Congress. U.S. Government Printing Office, Washington, D.C. 1974.

1630. U.S. Congress. Senate. Committee on Interior and Insular Affairs. SENATE'S NATIONAL FUELS AND ENERGY POLICY STUDY: PUBLICATIONS LIST. 93rd Congress. U.S. Government Printing Office, Washington, D.C. 1974.

1631. U.S. Congress. Senate. Committee on Interior and Insular Affairs. STATUS OF S. RES. 45: A NATIONAL FUELS AND ENERGY POLICY STUDY. 93rd Congress. U.S. Government Printing Office, Washington, D.C. 1974.

1632. U.S. Congress. Senate. Committee on Public Works. TOWARD A NATIONAL MATERIALS POLICY. U.S. Government Printing Office, Washington, D.C. 1969.

1633. U.S. Congress. Senate. Committee on Public Works. PROBLEMS AND ISSUES OF A NATIONAL MATERIALS POLICY. U.S. Government Printing Office, Washington, D.C. 1970.

1634. U.S. Congress. Senate. Select Committee on National Water Resources. WATER RESOURCES ACTIVITIES IN THE UNITED STATES: ELECTRIC POWER IN RELATION TO THE NATION'S WATER RESOURCES. U.S. Government Printing Office, Washington, D.C. 1960.

1635. U.S. Council on Economic Priorities. PETROLEUM INDUSTRY OVERVIEW. U.S. Council on Economic Priorities. Washington, D.C. 1970.

1636. U.S. Council on Environmental Quality. ENERGY AND THE ENVIRONMENT. U.S. Government Printing Office, Washington, D.C. n.d.

1637. U.S. Council on Environmental Quality. ENERGY AND ELECTRIC POWER.
       U.S. Government Printing Office, Washington, D.C. 1973.

1638. U.S. Council on Environmental Quality. Department of Commerce and
       Environmental Protection Agency. THE ECONOMIC IMPACT OF POLLU-
       TION CONTROL. U.S. Government Printing Office, Washington,
       D.C. 1972.

1639. U.S. Department of Agriculture. POWER TO PRODUCE: THE YEARBOOK
       OF AGRICULTURE, 1960. U.S. Government Printing Office,
       Washington, D.C. 1960.

1640. U.S. Department of Agriculture. Task Force on Spatial Heterogeneity
       in Agricultural Landscapes and Enterprises. MONOCULTURE IN
       AGRICULTURE: EXTENT, CAUSES, AND PROBLEMS. U.S. Department
       of Agriculture, Washington, D.C. 1973.

1641. U.S. Department of Commerce. PETROLEUM SUPPLY AND DEMAND IN THE
       FREE WORLD. National Technical Information Service, Spring-
       field, Virginia. n.d.

1642. U.S. Department of Commerce. ENERGY CONSERVATION AND THE BUSINESS
       COMMUNITY. U.S. Department of Commerce, Washington, D.C. 1973.

1643. U.S. Department of Commerce. ESTIMATED INTERNATIONAL FLOW OF
       PETROLEUM AND TANKER UTILIZATION. National Technical Informa-
       tion Service, Springfield, Virginia. 1973.

1644. U.S. Department of Commerce. Business and Defense Services Adminis-
       tration. ECONOMIC IMPACT OF AIR POLLUTION CONTROLS ON THE
       SECONDARY NONFERROUS METALS INDUSTRY. U.S. Government Printing
       Office, Washington, D.C. 1969.

1645. U.S. Department of Commerce. Office of Energy Programs. HOW TO
       START AN ENERGY MANAGEMENT PROGRAM. U.S. Government Printing
       Office, Washington, D.C. 1973.

1646. U.S. Department of Housing and Urban Development. BE AN ENERGY
       MISER IN YOUR HOME. U.S. Government Printing Office, Washington,
       D.C. 1973.

1647. U.S. Department of the Interior. SURFACE MINING AND OUR ENVIRONMENT.
       U.S. Government Printing Office, Washington, D.C. 1967.

1648. U.S. Department of the Interior. PROSPECTS FOR OIL-SHALE DEVELOPMENT:
       COLORADO, UTAH, AND WYOMING. U.S. Government Printing Office,
       Washington, D.C. 1968.

1649. U.S. Department of the Interior. ENVIRONMENTAL CRITERIA FOR ELECTRIC
       TRANSMISSION SYSTEMS. U.S. Government Printing Office, Washing-
       ton, D.C. 1970.

-119-

1650. U.S. Department of the Interior. U.S. ENERGY--A GENERAL REVIEW. U.S. Government Printing Office, Washington, D.C. 1971.

1651. U.S. Department of the Interior. U.S. ENERGY--A GENERAL REVIEW. U.S. Government Printing Office, Washington, D.C. 1972.

1652. U.S. Department of the Interior. U.S. ENERGY THROUGH THE YEAR 2000. U.S. Government Printing Office, Washington, D.C. 1972.

1653. U.S. Department of the Interior. FINAL ENVIRONMENTAL STATEMENT FOR PROTOTYPE OIL-SHALE LEASING PROGRAM. Volume 2, ENERGY ALTERNATIVES. U.S. Government Printing Office, Washington, D.C. 1973.

1654. U.S. Department of the Interior. Bureau of Mines. STRIPPABLE RESERVES OF BITUMINOUS COAL AND LIGNITE IN THE U.S. U.S. Government Printing Office, Washington, D.C. 1971.

1655. U.S. Department of the Interior. Coal Research Office. CLEAN ENERGY FROM COAL TECHNOLOGY. U.S. Government Printing Office, Washington, D.C. 1973.

1656. U.S. Department of the Interior. Defense Electric Power Administration. EMERGENCY PREPAREDNESS PROGRESS IN THE ELECTRIC UTILITY INDUSTRY. U.S. Government Printing Office, Washington, D.C. 1973.

1657. U.S. Department of the Interior. Office of Oil and Gas. THE MIDDLE EAST PETROLEUM EMERGENCY OF 1967. U.S. Government Printing Office, Washington, D.C. 1970.

1658. U.S. Department of State. CURRENT FOREIGN POLICY: ENERGY, COOPERATIVE WORLD ACTION TO SOLVE SHORTAGES. U.S. Government Printing Office, Washington, D.C. 1973.

1659. U.S. Energy Study Group. ENERGY R & D AND NATIONAL PROGRESS: A STUDY PREPARED FOR THE INTERDEPARTMENTAL STUDY GROUP. U.S. Government Printing Office, Washington, D.C. 1965.

1660. U.S. Energy Study Group. ENERGY R & D AND NATIONAL PROGRESS. U.S. Government Printing Office, Washington, D.C. 1973.

1661. U.S. Environmental Protection Agency. Office of Air and Water Programs. Office of Mobile Source Air Pollution Control. REPORT ON AUTOMOTIVE FUEL ECONOMY. U.S. Government Printing Office, Washington, D.C. 1973.

1662. U.S. Environmental Protection Agency and The Institute on Man and Science. ASSESSING THE SOCIAL IMPACTS OF OIL SPILLS: BACKGROUND PAPERS AND CONFERENCE PROCEEDINGS OF AN INVITATIONAL SYMPOSIUM, September 25-28, 1973, Rensselaerville, New York. The Institute on Man and Science, Rensselaerville, New York. February 1974.

1663. U.S. Federal Council for Science and Technology. Committee on Natural Resources. RESEARCH AND DEVELOPMENT ON NATURAL RESOURCES. U.S. Government Printing Office, Washington, D.C. 1963.

1664. U.S. Federal Power Commission. HYDROELECTRIC POWER RESOURCES OF THE UNITED STATES. U.S. Government Printing Office, Washington, D.C. 1968.

1665. U.S. Federal Power Commission. STATISTICS OF PRIVATELY OWNED ELECTRIC UTILITIES. U.S. Government Printing Office, Washington, D.C. 1970.

1666. U.S. Federal Power Commission. THE 1970 NATIONAL POWER SURVEY. U.S. Government Printing Office, Washington, D.C. 1971.

1667. U.S. Federal Power Commission. Bureau of Natural Gas. A STAFF REPORT ON NATIONAL GAS SUPPLY AND DEMAND. Staff Report Number 1. Bureau of Natural Gas, U.S. Federal Power Commission, Washington, D.C. 1969.

1668. U.S. Federal Power Commission. Bureau of Natural Gas. NATIONAL GAS SUPPLY AND DEMAND 1971-1990. Staff Report Number 2. U.S. Government Printing Office, Washington, D.C. 1972.

1669. U.S. General Accounting Office. OPPORTUNITY FOR BENEFITS THROUGH INCREASED USE OF COMPETITIVE BIDDING ON AWARD OIL AND GAS LEASES ON FEDERAL LANDS. U.S. Government Printing Office, Washington, D.C. 1970.

1670. U.S. General Accounting Office. HOW THE FEDERAL GOVERNMENT PARTICIPATES IN ACTIVITIES AFFECTING THE ENERGY RESOURCES OF THE UNITED STATES. U.S. Government Printing Office, Washington, D.C. 1973.

1671. U.S. General Accounting Office. ENERGY INFORMATION NEEDS. Study for the Committee on Interior and Insular Affairs. U.S. Senate. U.S. Government Printing Office, Washington, D.C. 1974.

1672. U.S. Industrial College of the Armed Forces. NATIONAL SECURITY MANAGEMENT: NATURAL AND ENERGY RESOURCES. U.S. Government Printing Office, Washington, D.C. 1960.

1673. U.S. Library of Congress. Environmental Policy Division. ENERGY-- THE ULTIMATE RESOURCE. Study submitted to the Task Force on Energy of the Subcommittee on Science, Resources, and Development. House Committee on Science and Astronautics. 92nd Congress. U.S. Government Printing Office, Washington, D.C. 1971.

-121-

1674. U.S. Library of Congress. Legislative Reference Service. Environmental Policy Division. THE ECONOMY, ENERGY, AND THE ENVIRONMENT: A BACKGROUND STUDY. Prepared for the use of the Joint Economic Committee. 91st Congress. U.S. Government Printing Office, Washington, D.C. 1970.

1675. U.S. Library of Congress. Science Policy Research Division. TOWARD A NATIONAL MATERIALS POLICY. U.S. Government Printing Office, Washington, D.C. 1969.

1676. U.S. National Bureau of Standards. TECHNICAL OPTIONS FOR ENERGY CONSERVATION IN BUILDINGS. U.S. Government Printing Office, Washington, D.C. 1973.

1677. U.S. National Industrial Pollution Control Council. Utilities Sub-Council. THE NATURAL GAS INDUSTRY AND THE ENVIRONMENT. U.S. Government Printing Office, Washington, D.C. 1972.

1678. U.S. Office of Emergency Preparedness. SURVEY OF ELECTRIC POWER PROBLEMS. U.S. Government Printing Office, Washington, D.C. 1971.

1679. U.S. Office of Emergency Preparedness. SURVEY OF ELECTRIC POWER PROBLEMS, SUMMER 1971. U.S. Government Printing Office, Washington, D.C. 1971.

1680. U.S. Office of Emergency Preparedness. THE POTENTIAL FOR ENERGY CONSERVATION: A STAFF STUDY. U.S. Government Printing Office, Washington, D.C. October 1972.

1681. U.S. Office of Science and Technology. FEDERAL RESEARCH AND DEVELOPMENT FOR CIVILIAN ENERGY PRODUCTION, CONVERSION AND TRANSMISSION. U.S. Government Printing Office, Washington, D.C. 1969.

1682. U.S. Petroleum Council. Committee on U.S. Energy Outlook. U.S. ENERGY OUTLOOK, A SUMMARY REPORT OF THE NATIONAL PETROLEUM COUNCIL. National Petroleum Council, Washington, D.C. 1972.

1683. Unnevehr, Clarence. "Is it Time to Review the Regulatory Process?" PUBLIC FORTNIGHTLY 84: 15-26. August 28, 1969.

1684. URS Research Corporation. GUIDELINES FOR GENERATING AND USING ELECTRIC POWER DURING PROLONGED EMERGENCIES. URS Research Corporation, San Mateo, California. 1971.

1685. Utility Executive Conference. PROCEEDINGS: BUILDING THE POWER SYSTEMS OF THE SEVENTIES. Utility Executive Conference, Chicago. 1971.

1686. Utton, Albert E. NATIONAL PETROLEUM POLICY: A CRITICAL REVIEW. University of New Mexico Press, Albuquerque. 1970.

1687. Utton, Albert E. "A Survey of National Laws on the Control of Pollution From Oil and Gas Operations on the Continental Shelf." COLUMBIA JOURNAL OF TRANSNATIONAL LAW 9: 331-361. Fall 1970.

1688. Vafai, John. "Market Demand Prorationing and Waste--A Statutory Confusion." ECOLOGY LAW QUARTERLY 2: 118-159. Winter 1972.

1689. Vahrman, Mark. "Energy in the Third World: Fuel and Power in Tanzania." ENERGY POLICY 2(2): 160-164. June 1974.

1690. Vansant, Carl. STRATEGIC ENERGY SUPPLY AND NATIONAL SECURITY. Praeger, New York. 1971.

1691. Vennard, Edwin. GOVERNMENT IN THE POWER BUSINESS. McGraw-Hill, New York. 1968.

1692. Vennard, Edwin. THE ELECTRIC POWER BUSINESS. McGraw-Hill, New York. 1970.

1693. Vernon, John M. PUBLIC INVESTMENT PLANNING IN CIVILIAN NUCLEAR POWER. Duke University Press, Durham, North Carolina. 1971.

1694. Vicker, Ray. THE KINGDOM OF OIL. Charles Scribners Sons, New York. 1974.

1695. Virden, Robert N. "Competition in the Natural Gas Transmission Industry." PUBLIC UTILITIES FORTNIGHTLY 84: 34-41. October 23, 1969.

1696. Vlachos, Evan. "The Energy Crisis: Despair Before Salvation." Department of Sociology, Colorado State University, Fort Collins. n.d.

1697. Von Neumann, John. "Can We Survive Technology?" FORTUNE 51: 106-108. June 1955.

1698. Vukadinovic, Radovan. "Oil and the US Mid-Eastern Policy." REVIEW OF INTERNATIONAL AFFAIRS 25(575): 10-12. March 20, 1974.

1699. Wagner, Aubrey J. "Power, Environment, and Your Pocketbook." PUBLIC UTILITIES FORTNIGHTLY 89(13): 27-31. June 22, 1972.

1700. Wagner, W.F. "Some Random Thoughts on Open Space, Environment, and Energy." ARCHITECTURAL RECORD. May 1972.

1701. Wagner, W.F. "Energy Conservation: A Potential Disaster Starts Getting the Attention it Deserves." ARCHITECTURAL RECORD. September 1972.

1702. Wagner, W.F. "Energy Conservation: Random Thoughts on Getting on With the Job." ARCHITECTURAL RECORD. July 1973.

1703. Wakefield, Stephen A. "Can Resources From Public Lands Be Developed in Environmentally Acceptable Ways?" In ENERGY AND THE ENVIRONMENT: A COLLISION OF CRISES, Irwin Goodwin (ed.). Publishing Sciences Group, Acton, Massachusetts. 1973.

1704. Walsh, John. "Vermont: A Power Deficit Raises Pressure for New Plants." SCIENCE 173: 1110-1115. September 17, 1971.

1705. Walsh, John. "Vermont: Forced to Figure in Big Power Picture." SCIENCE 174: 44-47. October 1, 1971.

1706. Walsh, John. "Britain and Energy Policy: Problems of Interdependence." SCIENCE 180: 1343-1347. June 29, 1973.

1707. Walsh, John. "Electric Power Research Institute: A New Formula for Industry R & D." SCIENCE 182: 263- . 1973.

1708. Walsh, John. "Problems of Expanding Coal Production." SCIENCE 184(4134): 336-339. April 19, 1974.

1709. Walsh, John and Abelson, Philip H. "The Executive: William E. Simon." Interview. SCIENCE 184(4134): 287-290.

1710. Walter, Norma. "Is There A Natural Gas Shortage?" EXCHANGE 31: 1-8. September 1970.

1711. Warman, H.R. "The Future of Oil." THE GEOGRAPHICAL JOURNAL 138(Part 3): 287-297. September 1972.

1712. Warner, Rawleigh, Jr. "That Alleged Oil Conspiracy." CONFERENCE BOARD RECORD 10(10): 10-15. October 1973.

1713. Warren, Frederick H. "Energy Policy and the Public." MILITARY ENGINEER 64: 239-241. August 1972.

1714. Washburn, Charles A. "Clean Water and Power." ENVIRONMENT 14(7): 40-44. September 1972.

1715. WASHINGTON LAW REVIEW. "Symposium: The Location of Electricity-Generating Facilities." (Power and the Environment--A Statutory Approach). WASHINGTON LAW REVIEW 47( ): 1- . 1971.

1716. Wasowski, Stanislaw. "The Fuel Situation in Eastern Europe." SOVIET STUDIES 21: 35-51. July 1969.

1717. Watt, Kenneth E.F. THE TITANIC EFFECT: PLANNING FOR THE UNTHINKABLE. E.P. Dutton, New York. 1974.

1718. Watt, Kenneth E.F. "Urban Land-Use Patterns, Energy Costs, and Pollution Production." Paper Presented at the Annual Meeting of the American Association for the Advancement of Science, San Francisco. February 1974.

1719. Watts, David. "Biogeochemical Cycles and Energy Flow in Environmental Systems." In PERSPECTIVES ON ENVIRONMENT, Ian R. Manners and Marvin W. Mikesell (eds.). Association of American Geographers, Washington, D.C. 1974 (Chapter 2).

1720. Weaver, Warren. "People, Energy and Food." SCIENTIFIC MONTHLY 78: 359-364. June 1954.

1721. Weidenfeld, Edward L. THIS NATION'S SUPPLY OF AND DEMAND FOR FUEL AND ENERGY RESOURCES. Committee on Interior and Insular Affairs, U.S. House of Representatives, Washington, D.C. 1972.

1722. Weinberg, Alvin M. "Social Institutions and Nuclear Energy." SCIENCE 177: 27-34. July 7, 1972.

1723. Weinberg, Alvin M. "Some Views of the Energy Crisis." AMERICAN SCIENTIST 61(1): 59-60. January-February 1973.

1724. Weinberg, Alvin M. and Hammond, R. Philip. "Limits to the Use of Energy." AMERICAN SCIENTIST 58: 412-418. July-August 1970.

1725. Weisberg, Barry. "Alaska--The Ecology of Oil." RAMPARTS 8(7): 25-33. January 1970. Reprinted in Glen A. Love and Rhonda M. Love (eds.), ECOLOGICAL CRISIS: READINGS FOR SURVIVAL. Harcourt Brace Jovanovich, New York. 1970. (pp. 187-203).

1726. Weiss, Brian E. "The Energetics of Cultural Maladaptation." Paper Presented at the Annual Meeting of the American Association for the Advancement of Science, San Francisco. February 1974.

1727. Wennergren, E. Boyd (ed.). ECONOMICS OF NATURAL RESOURCE DEVELOPMENT IN THE WEST: CURRENT PROBLEMS IN NATURAL RESOURCE USE. Report Number 4. Committee on the Economics of Natural Resources Development, Western Agricultural Economics Research Council, Utah State University, Logan. 1973.

1728. Werner, Morris Robert and Starr, John. TEAPOT DOME. Viking Press, New York. 1959.

1729. Wheatley, Charles F., Jr. "Natural Gas: Crisis in Regulation." PUBLIC POWER 29: 20-23. March-April 1971.

1730. White, David C. SUMMARY OF ENERGY STUDIES AND PROBLEMS. M.I.T. Press, Cambridge, Massachusetts. 1971.

1731. White, David C. "Energy, the Economy, and the Environment." TECHNOLOGY REVIEW 74(1): 18-31. October-November 1971.

1732. White, David C. The U.S. Energy Crisis: A Scientist's View." ENERGY POLICY 1(2): 130-135. September 1973.

1733. White, Donald E. "Geothermal Energy." In ENERGY, THE ENVIRONMENT, AND HUMAN HEALTH, Asher J. Finkel (ed.). Publishing Sciences Group, Acton, Massachusetts. 1973.

1734. White, Irvin L. "Energy Policy-Making: Limitations of a Conceptual Model." In THE ENERGY CRISIS, Richard S. Lewis and Bernard I. Spinrod (eds.). Educational Foundation for Nuclear Science, Chicago. 1972.

1735. White, L. "Environmental Concern is a Major Problem." PUBLIC POWER. May 1970.

1736. White, Lee. "The Consumer's Slant on the Energy Crisis." In ENERGY AND THE ENVIRONMENT: A COLLISION OF CRISES, Irwin Goodwin (ed.). Publishing Sciences Group, Acton, Massachusetts. 1973.

1737. White, Leslie A. "Energy in the Development of Civilization." IMPACT OF SCIENCE ON SOCIETY 1: 38-40. July-September 1950.

1738. White, Leslie A. "The Energy Theory of Cultural Development." In GHURYE FELICITATION VOLUME, K.M. Kapadia (ed.). Popular Book Depot, Bombay. (pp. 1-10).

1739. White, Lynn T., Jr. MEDIEVAL TECHNOLOGY AND SOCIAL CHANGE. Clarenden Press, Oxford. 1962.

1740. White, Warren. "Impacts of Northern Great Plains Coal Related Development on Nebraska." State Office of Planning and Programming, Lincoln, Nebraska. April 1974.

1741. Whittenmore, F. Case. "How Much is Reserve?" ENVIRONMENT 15(7): 16-35. September 1973.

1742. Williamson, Harold F., et al. THE AMERICAN PETROLEUM INDUSTRY. 2 Volumes. Northwestern University Press, Evanston, Illinois. 1959, 1963.

1743. Willrich, Mason. CIVIL NUCLEAR POWER AND INTERNATIONAL SECURITY. Praeger, New York. 1971.

1744. Willrich, Mason. "The Energy-Environment Conflict: Siting Electric Power Facilities." VIRGINIA LAW REVIEW 58: 257-336. February 1972.

1745. Willrich, Mason and Taylor, Theodore B. NUCLEAR THEFT: RISKS AND SAFE-GUARDS. Ballinger Publishing Co., Cambridge, Massachusetts. 1974.

1746. Wilpers, John. "The Energy Crisis: Who's in Charge Here?" GOVERNMENT EXECUTIVE. February 1973.

1747. Wilpers, John. "New White Knight and the Energy Dragon." GOVERNMENT EXECUTIVE 5(8): 44-50. August 1973.

1748. Wilson, Carroll L. "A Plan for Energy Independence." FOREIGN AFFAIRS 51(4): 657-675. July 1973.

1749. Wilson, George. "The Effect of Rate Regulation on Resource Transportation." In THE CRISIS OF THE REGULATORY COMMISSION. W.W. Norton, New York. 1970.

1750. Wilson, John W. "Residential Demand for Electricity." QUARTERLY REVIEW OF ECONOMICS AND BUSINESS 11: 7-22. Spring 1971.

1751. Wilson, Mitchell. ENERGY. Life Science Library. Time, Inc. New York. 1967.

1752. Wilson, Richard. "Power Policy: Plan or Panic?" BULLETIN OF THE ATOMIC SCIENTISTS 28(5): 29-30. May 1972.

1753. Wilson, Richard. "Natural Gas Is a Beautiful Thing?" SCIENCE AND PUBLIC AFFAIRS BULLETIN OF THE ATOMIC SCIENTISTS 29(7): 35-40. September 1973.

1754. Wilson, W. and Jones, R. ENERGY, ECOLOGY, AND THE ENVIRONMENT. Academic Press, New York. (Forthcoming).

1755. Winger, John G., et al. FUTURE GROWTH OF THE WORLD PETROLEUM INDUSTRY. Chase Manhattan Bank, New York. 1961.

1756. Winger, John G., Emerson, John D. and Gunning, Gerald B. OUTLOOK FOR ENERGY IN THE UNITED STATES. Chase Manhattan Bank, New York. 1968.

1757. Winsche, W.E., Hoffman, K.C. and Salzano, F.J. "Hydrogen: Its Future Role in the Nation's Energy Economy." SCIENCE 180: 1325-1332. June 29, 1973.

1758. Winter, Thomas C. "CEQ and Its Role in Environmental Policy." In ENVIRONMENTAL IMPACT ANALYSIS: PHILOSOPHY AND METHODS, Robert B. Ditton and Thomas L. Goodale (eds.). Sea Grant Publications Office, University of Wisconsin, Madison. 1972.

1759. Wirtz, Richard S. "Electric-Utility Interconnections: Power to the People." STANFORD LAW REVIEW 21: 1714-1733. June 1969.

1760. Witt, Matt. "The New Energy Barons: How Big Oil Controls the Coal Industry." UNITED MINE WORKERS JOURNAL (12): 4-7. July 15-31, 1973.

1761. Witteveen, Johannes. "Energy and Money." EUROPEAN COMMUNITY (175): 15-16. April 1974.

1762. Wohlwill, Joachim F. "Adaptation, Adjustment, Attitude, and Energy." Paper Presented at the Annual Meeting of the American Association for the Advancement of Science, San Francisco. February 1974.

1763. Wolff, Anthony. "The Price of Power." HARPER'S MAGAZINE 244: 36-38. May 1972.

1764. Woodbury, Angus M. "Colorado Dam Controversy." SCIENTIFIC MONTHLY 82: 304-313. June 1956.

1765. Wright, James H. "The Future of Electric Energy." PUBLIC UTILITIES FORTNIGHTLY 90: 15-19. December 21, 1972.

1766. Yager, Joseph and Steinberg, Eleanor. ENERGY AND U.S. FOREIGN POLICY. Ford Foundation. n.d.

1767. YALE LAW JOURNAL. "Gasoline Marketing and the Robinson-Patman Act." YALE LAW JOURNAL 82(8): 1706-1718. July 1973.

1768. Yee, J.E. OIL POLLUTION OF MARINE WATERS. Bibliography No. 6. U.S. Department of the Interior Library. U.S. Government Printing Office, Washington, D.C. November 1967.

1769. Young, H.J. "Environmental Impacts From Power Plant Sitings and Distribution of Energy." In ENERGY, THE ENVIRONMENT, AND HUMAN HEALTH, Asher J..Finkel (ed.). Publishing Sciences Group, Acton, Massachusetts. 1973.

1770. Young, H.J. "Power Plant Siting and the Environment." OKLAHOMA LAW REVIEW 26( ): 193- . 1973.

1771. Zachariesen, F. OIL POLLUTION IN THE SEA: PROBLEMS FOR FUTURE WORK. Research Paper Number P-432. Institute for Defense Analysis, Arlington, Virginia. 1968.

1772. Zaffarane, R.F., et al. SUPPLY AND DEMAND FOR ENERGY IN THE UNITED STATES BY STATES AND REGIONS, 1960 AND 1965. Information Circular Number 8434. U.S. Bureau of Mines. U.S. Government Printing Office, Washington, D.C. 1969.

1773. ZoBell, Claude E. "The Occurrence, Effects and Fate of Oil Polluting the Sea." INTERNATIONAL JOURNAL OF AIR AND WATER POLLUTION 7( ): 173-193. 1962.

1774. Zraket, Charles A. "The Global Energy-Environment Problem." In ENERGY AND THE ENVIRONMENT: A COLLISION OF CRISES, Irwin Goodman (ed.). Publishing Sciences Group, Acton, Massachusetts. 1973.

1775. Zybenko, Roman. "Fuel and Power Resources." STUDIES ON THE SOVIET UNION 8(1): 9-13. 1968.

ADDENDUM

ADDENDUM

1776. Abelson, Philip. "Dealing Now with the Energy Crisis." CURRENT 150: 56-57. April 1973.

1777. Adelman, Morris Albert, et al. "Energy Self Sufficiency: An Economic Evaluation." TECHNOLOGY REVIEW 76(6): 44-47, 56-58. May 1974.

1778. Aldrich, James L., et al. (eds.). ENERGY, ENVIRONMENT AND EDUCATION: A WORKING PAPER. Conservation Foundation, Washington, D.C. 1973.

1779. Alfven, H. "Energy and Environment." BULLETIN OF THE ATOMIC SCIENTISTS: 5. May 1972.

1780. Allais, M. "Method of Appraising Economic Prospects of Mining Exploration Over Large Territories." MANAGEMENT SCIENCE 3: 285-345. July 1957.

1781. ALTERNATIVES MAGAZINE. "Energy: Bibliography Number 7." ALTERNATIVES MAGAZINE. 1974.

1782. American Institute of Architects. "Energy Conservation in Building Design." Draft Report to the Energy Policy Project. Ford Foundation, New York. May 1974.

1783. Anderson, Kent P. TOWARD ECONOMETRIC ESTIMATION OF INDUSTRIAL ENERGY DEMAND: AN EXPERIMENTAL APPLICATION TO THE PRIMARY METALS INDUSTRY. R-719-NSF. Rand Corporation, Santa Monica, California. 1971.

1784. Associated Universities, Inc. REFERENCE ENERGY SYSTEMS AND RESOURCE DATA FOR USE IN THE ASSESSMENT OF ENERGY TECHNOLO- GIES. U.S. Government Printing Office, Washington, D.C. 1972.

1785. Austin, Richard Cartwright and Borrelli, Peter. THE STRIP MINING OF AMERICA: AN ANALYSIS OF SURFACE COAL MINING AND THE ENVIRONMENT. Sierra Club, New York. 1971.

1786. Baldwin, Malcolm F. PUBLIC POLICY ON OIL: AN ECOLOGICAL PER- SPECTIVE. Reprinted from ECOLOGY LAW QUARTERLY. Conserva- tion Foundation, Washington, D.C. 1974.

1787. Balestra, Pietro. THE DEMAND FOR NATURAL GAS IN THE UNITED STATES. Harvard University Press, Cambridge, Massachusetts. 1966.

1788. Ball, R.H., et al. CALIFORNIA'S ELECTRICITY QUANDARY: II. PLANNING FOR POWER PLANT SITING. Rand Corporation, Santa Monica, California. 1972.

1789. Baughman, Martin L. DYNAMIC ENERGY SYSTEM MODELING--INTERFUEL COMPETITION. Report Number MIT-EL-72-1. Energy Analysis and Planning Group, School of Engineering, Massachusetts Institute of Technology, Cambridge, Massachusetts. 1972.

1790. Baxter, R.E. and Rees, R. "An Analysis of the Industrial Demand for Electricity." ECONOMIC JOURNAL 78(310): 277-298. June 1968.

1791. Beak Consultants Limited. GAS AND OIL PIPELINES IN THE MACKENZIE VALLEY AND NORTHERN YUKON: CONSIDERATIONS FOR CONTINGENCY PLANNING. Public Information, Indian and Northern Affairs, Ottawa. 1974.

1792. Beck, Robert E. and Johnson, Jerome E. LEGAL ASPECTS OF COAL LEASING AND SALES, AND STRIP MINING RECLAMATION IN NORTH DAKOTA. Extension Bulletin Number 22. Cooperative Extension Service, North Dakota State University, Fargo. December 1973.

1793. Benn, T. "Technology Assessment and Political Power." NEW SCIENTIST V(58): 487-490. 1973.

1794. Bergman, P.D., et al. AN ECONOMIC EVALUATION OF MHD-STEAM POWER PLANTS EMPLOYING COAL GASIFICATION. Report of Investigations 7796. Bureau of Mines, U.S. Department of the Interior, Washington, D.C. 1973.

1795. Bissett, D. THE LOWER MACKENZIE REGION: AN AREA ECONOMIC SURVEY. Northern Administration Branch, Indian and Northern Affairs, Ottawa. 1967.

1796. Boeking, Dick. "The Energy Crisis: A Time to Choose." In ENERGY AND THE ENVIRONMENT, Ian E. Efford and Barbara M. Smith (eds.). Institute of Resource Ecology, University of British Columbia, Vancouver. 1972. (pp. 176-190).

1797. Bohi, Douglas R. and Russell, Milton. "The Energy 'Crisis" as a Problem in Economic Adjustment." In ENERGY AND AGRICULTURE: RESEARCH IMPLICATIONS, Loyd Fischer and Arlo Biere (eds.). North Central Research Strategy Committee on Natural Resource Development. October 1973. (pp. 1-40).

1798. Bowman, Mary Jane and Flaynes, W.W. RESOURCES AND PEOPLE IN WESTERN KENTUCKY. Johns Hopkins Press, Baltimore, Maryland. 1963.

-131-

1799. Box, Thadis W. "The Energy Crisis and the Fate of Strip Mine Lands." UTAH SCIENCE 34(4): 117-120. December 1973.

1800. Brannon, Gerald M. ENERGY TAXES AND SUBSIDIES. Report to the Ford Foundation Energy Policy Project. Ballinger, Cambridge, Massachusetts. 1974.

1801. Brannon, Gerald M. (ed.). STUDIES IN ENERGY TAX POLICY. Report to the Ford Foundation Energy Policy Project. Ballinger, Cambridge, Massachusetts. 1974.

1802. British Petroleum. BP STATISTICAL REVIEW OF THE WORLD OIL INDUSTRY, 1973. British Petroleum, London. 1973.

1803. Brock, Samuel M. and Brooks, David B. THE MYLES JOB MINE--A STUDY OF BENEFITS AND COST OF SURFACE MINING FOR COAL IN NORTHERN WEST VIRGINIA. Research Series 1. Appalachian Center, West Virginia University, Morgantown. 1968.

1804. Brookings Institute. "Energy and U.S. Foreign Policy." Draft Report to the Energy Policy Project. Ford Foundation, New York. June 1974.

1805. Brooks, David B. "Strip Mine Reclamation and Economic Analysis." NATURAL RESOURCES JOURNAL 6: 13-44. 1966.

1806. Brooks, David B. "Strip Mining, Reclamation, and the Public Interest." AMERICAN FORESTS 72: 51-57. 1966.

1807. Brown, Keith C. (ed.). REGULATION OF THE NATURAL GAS PRODUCING INDUSTRY. Resources for the Future, Inc., Washington, D.C. 1970.

1808. Burnham, James. "Energy: The Strategic Dimension: The Protracted Conflict." NATIONAL REVIEW: 412. April 13, 1973.

1809. Callahan, John C. and Callahan, Jacqueline C. EFFECTS OF STRIP MINING AND TECHNOLOGICAL CHANGE ON COMMUNITIES AND NATURAL RESOURCES IN INDIANA'S COAL MINING REGION. Research Bulletin Number 871. Agricultural Experiment Station, Purdue University, Lafayette, Indiana. 1971.

1810. Canada. Environment Protection Board. TOWARDS AN ENVIRONMENTAL IMPACT ASSESSMENT OF THE PORTION OF THE MACKENZIE GAS PIPE-LINE FROM ALASKA TO ALBERTA. Environment Protection Board, Winnipeg. 1973.

1811. Canada. Federal Government. AN ENERGY POLICY FOR CANADA: PHASE 1. The Queen's Printer, Ottawa, Canada. 1973.

1812. Canada, Indian and Northern Affairs, Northern Economic Development Branch and MPS Associates, Ltd. REGIONAL IMPACT OF A NORTHERN GAS PIPELINE. Information Canada, Ottawa. Volumes 1-7. 1973-1974.

1813. Carlson, Albert S. ECONOMIC GEOGRAPHY OF INDUSTRIAL MATERIALS. Reinhold Publishing Company, New York. 1956.

1814. Carter, Ann P. "Application of Input/Output Analysis to Energy Problems." SCIENCE: 325-329. April 19, 1974.

1815. Carter, Ann P. (ed.). STRUCTURAL INTERDEPENDENCE, ENERGY AND THE ENVIRONMENT. University Press of New England, Hanover, New Hampshire. (Forthcoming).

1816. Catton, William R., Jr. "Extensional Orientation and the Energy Problem." ETC: A REVIEW OF GENERAL SEMANTICS 30(4): 344-356. December 1973.

1817. Caudill, Harry Monroe. "Strip Mining: Partnership in Greed." AMERICAN FORESTS 79: 16-19. May 1973.

1818. Caudill, Harry Monroe. "Farming and Mining." ATLANTIC MONTHLY 232(3): 85-90. September 1973.

1819. Chaplin, Ronald L. SPATIAL CHANGES IN COAL EMPLOYMENT WITHIN SOUTHERN ILLINOIS 1900-1960. Unpublished M.A. thesis. Southern Illinois University, Carbondale. 1961.

1820. Charle, Edwin G., Jr. THE DEMAND FOR POWER GENERATION IN THE TENNESSEE VALLEY AND THE IMPACT OF CHANGING DEMAND PATTERNS ON A SUPPLYING COAL FIELD. Unpublished Ph.D. dissertation, Indiana University, Bloomington. 1958.

1821. Chase, Nina Ross. A BENEFIT-COST ANALYSIS OF OHIO'S RECLAIMED COAL LANDS. Unpublished M.A. thesis, Ohio State University, Columbus. 1967.

1822. Cinq-Mars, J. PRELIMINARY ARCHAEOLOGICAL STUDY, MACKENZIE CORRIDOR. Report Number 73-10. Information Canada, Ottawa. 1973.

1823. Clark, Wilson. ENERGY FOR SURVIVAL: THE ALTERNATIVE TO EXTINCTION. Anchor Press, New York. 1974.

1824. Clarks, Ronald O. and List, Peter C. (eds.). ENVIRONMENTAL SPECTRUM. D. Van Nostrand, New York. 1974.

1825. Cochran, T.B. THE LIQUID METAL FAST BREEDER REACTOR: AN ENVIRONMENTAL AND ECONOMIC CRITIQUE. Resources for the Future, Washington, D.C. 1974.

1826. Commoner, Barry, Boksenbaum, Howard and Corr, Michael (eds.). ENERGY AND HUMAN WELFARE: A CRITICAL ANALYSIS. Macmillan, New York. 1974.

1827. Conaway, James. "The Last of the West: Hell, Strip It!" ATLANTIC MONTHLY 232(3): 91-103. September 1973.

1828. Conference Board. ENERGY CONSUMPTION IN MANUFACTURING. Report to the Ford Foundation Energy Policy Project. Ballinger, Cambridge, Massachusetts. 1974.

1829. Conrad, Jon M. UNCERTAIN EXTERNALITY: THE CASE OF OIL POLLUTION. Unpublished Ph.D. dissertation. University of Wisconsin, Madison. 1973.

1830. Conservation Foundation. THE SOUTHWEST ENERGY COMPLEX: A POLICY EVALUATION. Conservation Foundation, Washington, D.C. 1974.

1831. Copeland, Otis L. and Parker, Paul E. "Land Use Aspects of the Energy Crisis and Western Mining." JOURNAL OF FORESTRY: 70. November 1972.

1832. Cottrell, W.F. "Death by Dieselization: A Case Study in the Reaction to Technological Change." AMERICAN SOCIOLOGICAL REVIEW 16(3). June 1951.

1833. Crump, Leslie. SUPPLY AND DEMAND FOR ENERGY IN THE U.S. BY STATES AND REGIONS, 1960 and 1965. U.S. Department of the Interior, Washington, D.C. 1969.

1834. Crump, Leslie H. and Yasnowsky, Phillip N. SUPPLY AND DEMAND FOR ENERGY IN THE U.S. BY STATES AND REGIONS, 1960 and 1965: PART 4, PETROLEUM AND NATURAL GAS LIQUIDS. Bureau of Mines Information Circular Number 8411. U.S. Department of the Interior, Washington, D.C. 1969.

1835. Cummings, R.G. "Some Extensions of the Economic Theory of Exhaustible Resources." WESTERN ECONOMIC JOURNAL 7: 201-210. September 1969.

1836. DAEDALUS. "The No-Growth Society." DAEDALUS 102: 4. Fall 1973.

1837. Dalsted, Norman L. and Leistritz, F. Larry. "North Dakota Coal Resources and Development Potential." NORTH DAKOTA FARM RESEARCH 31(6): 3-11. July-August 1974.

1838. Dalsted, Norman L. and Leistritz, F. Larry. A SELECTED BIBLIOGRAPHY ON COAL-ENERGY DEVELOPMENT OF PARTICULAR INTEREST TO THE WESTERN STATES. Miscellaneous Report Number 16. Department of Agricultural Economics, North Dakota Agricultural Experiment Station, North Dakota State University, Fargo. April 1974.

1839. Daly, H.E. (ed.). TOWARD A STEADY STATE ECONOMY. W.H. Freeman, San Francisco. 1973.

1840. Darmstadter, Joel. "Appendix: Energy Consumption: Trends and Patterns." In ENERGY, ECONOMIC GROWTH, AND THE ENVIRONMENT, Sam H. Schurr (ed.). Published for Resources for the Future by the Johns Hopkins Press, Baltimore. 1972.

1841. Data Resources Inc. "Energy Projections: An Economic Model." Draft Report to the Energy Policy Project. Ford Foundation, New York. May 1974.

1842. David, Edward E., Jr. "Energy: A Strategy of Diversity." TECHNOLOGY REVIEW 25(7): 26-31. June 1973.

1843. Davis, Jack. "Energy and the Environment: The Federal Point of View." In ENERGY AND THE ENVIRONMENT, Ian E. Efford and Barbara M. Smith (eds.). Institute of Resource Ecology, University of British Columbia, Vancouver. 1972. (pp. 191-205).

1844. Deasy, George F. "Coal Strip Pits in the Northern Appalachian Landscape." THE JOURNAL OF GEOGRAPHY 58: 72-81. 1959.

1845. de Carmoy, Guy, et al. COOPERATIVE APPROACHES TO WORLD ENERGY PROBLEMS. Brookings Institute, Washington, D.C. 1974.

1846. Denis, Sylvain. SOME ASPECTS OF THE ENVIRONMENT AND ELECTRIC POWER GENERATION. P-4777. Rand Corporation, Santa Monica, California. 1972.

1847. Depree, Walter, Jr. and West, James A. "United States Energy Through the Year 2000." U.S. Department of the Interior, Washington, D.C. 1972.

1848. Dinkel, R. Michael and Guernsey, Lee. "An Economic Appraisal of Reclamation Practices on a Strip Coal Mine Site in Greene County, Indiana." INDIANA ACADEMY OF SCIENCE 78: 355-362. 1968.

1849. Doctor, R.D. THE GROWING DEMAND FOR ENERGY. P-4759. Rand Corporation, Santa Monica, California. 1972.

1850. Doerr, Arthur M. "Coal Mining and Changing Land Patterns in Oklahoma." LAND ECONOMICS 38: 51-56. 1962.

1851. Dole, S.H. and Papetti, P.A. ENVIRONMENTAL FACTORS IN THE PRODUCTION AND USE OF ENERGY. Rand Corporation, Santa Monica, California. 1973.

1852. Dregne, Harold E. (ed.). ARID LANDS IN TRANSITION. Publication Number 90. American Association for the Advancement of Science, Washington, D.C. 1970.

1853. Duchesneau, Thomas D. "Competition in the Energy Industry." Draft Report to the Energy Policy Project. Ford Foundation, New York. 1974.

1854. Eastman, Clyde. "Socio-economic Characteristics Related to Environmental Concern: The Case of the Four Corners Electric Power Complex." Department of Agricultural Economics and Agricultural Business, New Mexico State University, Las Cruces, New Mexico. 1974.

1855. Economic Staff Group. REGIONAL IMPACT OF A NORTHERN GAS PIPELINE: VOLUME I-SUMMARY. Information Canada, Ottawa. 1973.

1856. Edwards, Corwin D. "Conglomerate Bigness as a Source of Power." In BUSINESS CONCENTRATION AND PRICE POLICY, Princeton University Press, Princeton. 1955. (pp. 331-360).

1857. Efford, Ian E. "Energy Addiction: A Social Disease." In ENERGY AND THE ENVIRONMENT, Ian E. Efford and Barbara M. Smith (eds.). Institute of Resource Ecology, University of British Columbia, Vancouver. 1972. (pp. 206-219).

1858. Ehrlich, Paul and Holdren, J.P. "The Heat Barrier." SATURDAY REVIEW. April 3, 1971.

1859. ENERGY POLICY (eds.). ENERGY MODELING. IPC Science and Technology Press, Guildford, England. 1974.

1860. Enke, Stephen. "Population Growth and Economic Growth." THE PUBLIC INTEREST 32. Summer 1973.

1861. Erickson, Edward W. and Spann, Robert M. "Supply Response in a Regulated Industry: The Case of Natural Gas." THE BELL JOURNAL OF ECONOMICS AND MANAGEMENT SCIENCE 2(1): 94-121. 1971.

1862. Faucett, Jack, Associates, Inc. PROJECT INDEPENDENCE AND ENERGY CONSERVATION: TRANSPORTATION SECTORS. U.S. Federal Energy Administration, Washington, D.C. 1974.

1863. Fischer, Loyd K. "Energy Crisis and the Role of Agriculture." NATURAL RESOURCES: 9-11. Spring 1973.

1864. Foell, Wesley K. "The Energy Source of the Future: Fast Breeder Reactor or ?" In ENERGY AND THE ENVIRONMENT, Ian E. Efford and Barbara M. Smith (eds.). Institute of Resource Ecology, University of British Columbia. 1972. (pp. 107-140).

1865. Ford, D.F. and Kendall, H.W. "Catastrophie Nuclear Accidents." In THE NUCLEAR FUEL CYCLE: A SURVEY OF THE PUBLIC HEALTH, ENVIRONMENTAL AND NATIONAL SECURITY EFFECTS OF NUCLEAR FUELS. Union of Concerned Scientists. 1973.

1866. Ford Foundation Energy Policy Project. A TIME TO CHOOSE: AMERICA'S ENERGY FUTURE. Ballinger Publishing Company, Cambridge, Massachusetts. 1974.

1867. Forth, T.C., et al. MACKENZIE VALLEY DEVELOPMENT: SOME IMPLICATIONS FOR PLANNERS. Report Number 73-45. Government of the Northwest Territories. Information Canada, Ottawa. 1974.

1868. Foster Associates, Inc. ENERGY PRICES 1960-73. Report to the Ford Foundation Energy Policy Project. Ballinger, Cambridge, Massachusetts. 1974.

1869. Foster, John S. and Stewart, Gordon C. "Questions About Canadian Nuclear Power." In ENERGY AND THE ENVIRONMENT, Ian E. Efford and Barbara M. Smith (eds.). Institute of Resource Ecology, University of British Columbia, Vancouver. 1972. (pp. 90-106).

1870. Foster, R. Bruce. "Projected Costs of Alternate Sources of Gas." In REGULATION OF THE NATURAL GAS PRODUCING INDUSTRY, Keith C. Brown (ed.). Resources for the Future, Washington, D.C. 1970. (pp. 63-88).

1871. Freeman, S. David. "New Policies for Energy and the Environment." In ENERGY AND THE ENVIRONMENT, Ian E. Efford and Barbara M. Smith (eds.). Institute of Resource Ecology, University of British Columbia, Vancouver. 1972. (pp. 32-44).

1872. Fried, J. "Urbanization and Ecology in the Canadian Northwest Territories." ARCTIC ANTHROPOLOGY 2(2). 1964.

1873. Frowley, Margaret L. SURFACE MINED AREAS: CONTROL AND RECLAMATION OF ENVIRONMENTAL DAMAGE. Bibliography Series Number 37. Office of Library Service, U.S. Department of the Interior, Washington, D.C. 1971.

1874. Gaffney, Mason (ed.). EXTRACTIVE RESOURCES AND TAXATION. University of Wisconsin Press, Madison. 1967.

1875. Gemini North. MACKENZIE VALLEY SOCIAL IMPACT STUDY. Prepared for the Government of the Northwest Territories. 1973.

1876. Gifford, Gerald F., Dwyer, Don D. and Norton, Brien E. A BIBLIOGRAPHY OF LITERATURE PERTINENT TO MINING RECLAMATION IN ARID AND SEMI-ARID ENVIRONMENTS. The Environment and Man Program, Utah State University, Logan. 1972.

1877. Gofman, John W. "Nuclear Electric Power: Do We Need It and How Should We Decide?" In ENERGY AND THE ENVIRONMENT, Ian E. Efford and Barbara M. Smith (eds.). Institute of Resource Ecology, University of British Columbia, Vancouver. 1972. (pp. 69-89).

1878. Gold, Raymond L., et al. "A Comparative Case Study of the Impact of Coal Development on the Way of Life of People in the Coal Areas of Eastern Montana and Northeastern Wyoming." Institute for Social Science Research, University of Montana, Missoula. June 1974.

1879. Goldberg, Michael A. "Energy Supply and Economic Growth: Some Costs, Doubts and Dangers." In ENERGY AND THE ENVIRONMENT, Ian E. Efford and Barbara M. Smith (eds.). Institute of Resource Ecology, University of British Columbia, Vancouver. 1972. (pp. 141-161).

1880. Gonzalez, Richard J. "Interfuel Competition for Future Energy Markets." JOURNAL OF THE INSTITUTE OF PETROLEUM 54(535). 1968.

1881. Goodland, Robert (ed.). POWER LINES AND THE ENVIRONMENT. Cary Arboretum of the New York Botanical Gardens, Millbrook, New York. 1973.

1882. Gordian Associates, Inc. THE POTENTIAL FOR ENERGY CONSERVATION IN NINE SELECTED INDUSTRIES. Gordian Associates, Inc. 1974.

1883. Gordon, R.L. "A Reinterpretation of the Pure Theory of Exhaustion." JOURNAL OF POLITICAL ECONOMY 75: 274-286. June 1967.

1884. Graham, H.D. THE ECONOMICS OF STRIP COAL MINING. Bulletin Number 66. Bureau of Economics and Business Research, University of Illinois, Urbana. 1948.

1885. Gray, D.M., et al. ENERGY BUDGET IN THE ARTIC. Information Canada, Ottawa. 1973.

1886. Gray, J. Lorne. "Nuclear Power: An Energy Source for Canada." In ENERGY AND THE ENVIRONMENT, Ian E. Efford and Barbara M. Smith (eds.). Institute of Resource Ecology, University of British Columbia, Vancouver. 1972. (pp. 45-68).

1887. Grenon, Michel. "Energy R & D in the Industrialized Countries." Draft Report to the Energy Policy Project. Ford Foundation, New York. 1974.

1888. Guernsey, Lee. "Land Use Changes Caused by a Quarter Century of Strip Coal Mining in Indiana." INDIANA ACADEMY OF SCIENCE 69: 200-209. 1959.

1889. Guernsey, Lee. "Settlement Changes Caused by Strip Coal Mining in Indiana." INDIANA ACADEMY OF SCIENCE 70: 158-164. 1960.

1890. Gwynn, Thomas A. "A Quality Environment Can Be Maintained While Strip Mining Montana's Coal." Paper Presented at Upper Missouri Water Users Conference. December 1972.

1891. Hamilton, Richard E. "Canada's 'Exportable Surplus' Natural Gas Policy: A Theoretical Analysis." LAND ECONOMICS 49(3): 251-259. 1973.

1892. Hammond, Allen L., Metz, William D. and Maugh, Thomas H., II. ENERGY AND THE FUTURE. American Association for the Advancement of Science, Washington, D.C. 1973.

1893. Hannah, H.W. and Vanderoliet, B. "Effects of Strip Mining on Agricultural Areas in Illinois and Suggested Remedial Measures." LAND ECONOMICS 15: 296-311. 1939.

1894. Hannon, Bruce, Herendeen, Robert and Sebald, Anthony. "The Energy Content of Certain Consumer Products." Draft Report to the Ford Foundation Energy Policy Project. Energy Research Group, Center for Advanced Computation, University of Illinois. July 1973.

1895. Harberger, A.C. "The Taxation of Mineral Industries." In FEDERAL TAX POLICY FOR ECONOMIC GROWTH AND STABILITY, Joint Committee on the Economic Report, 84th Congress, 1st Session. U.S. Government Printing Office, Washington, D.C. 1955. (pp. 439-449).

1896. Harl, Neil E. "An Overview." In ENERGY AND AGRICULTURE: RESEARCH IMPLICATIONS, Loyd Fischer and Arlo Biere (eds.). North Central Research Strategy Committee on Natural Resource Development. October 1973.

1897. Harvard University. Department of Landscape Architecture. THREE APPROACHES TO ENVIRONMENTAL RESOURCE ANALYSIS. Conservation Foundation, Washington, D.C. 1974.

1898. Harvey, Douglas G. and Menchen, W. Robert. THE AUTOMOBILE, ENERGY AND THE ENVIRONMENT: A TECHNOLOGY ASSESSMENT OF ADVANCED AUTOMOTIVE PROPULSION SYSTEMS. Hittman Associates, Columbia, Maryland. 1974.

1899. Hass, Jerome E., Mitchell, Edward J. and Stone, Bernell K. FINANCING THE ENERGY INDUSTRY. A Report to the Ford Foundation Energy Policy Project. Ballinger, Cambridge, Massachusetts. 1974.

1900. Heginbottom, J.A. EFFECTS OF SURFACE DISTURBANCE ON PERMAFROST. Information Canada, Ottawa. 1973.

1901. Henderson, James N. THE EFFICIENCY OF THE COAL INDUSTRY. Harvard University Press, Cambridge, Massachusetts. 1958.

1902. Henry, J.P., Jr. and Louks, B.M. "An Economic Study of Pipeline Gas Production from Coal." CHEMICAL TECHNOLOGY 1: 238-247. April 1971.

1903. Hertsgaard, Thor A. and Leistritz, F. Larry. "Environmental Impact of Strip Mining: The Economic and Social Viewpoints." In SOME ENVIRONMENTAL ASPECTS OF STRIP MINING IN NORTH DAKOTA, Mohan K. Wali (ed.). Educational Series 5. North Dakota Geological Survey, Grand Forks. 1974. (pp. 73-86).

1904. Higgins, G. THE LOWER MACKENZIE REGION: AN AREA ECONOMIC SURVEY. Northern Administration Branch, Indian and Northern Affairs, Ottawa. 1969.

1905. Hill, Jack K. SOCIAL AND ECONOMIC IMPLICATIONS OF STRIP MINING IN HARRISON COUNTY. Unpublished M.A. thesis. Ohio State University, Columbus. 1965.

1906. Hirst, Eric. "The Energy Crisis: Energy Versus Environment." THE LIVING WILDERNESS: Winter 1972-1973. Reprinted CURRENT 150: 50-56. April 1973.

1907. Hirst, Eric. ENERGY USE FOR FOOD IN THE UNITED STATES. Oak Ridge National Laboratory, Oak Ridge, Tennessee. 1973.

1908. Hittman Associates. RESIDENTIAL ENERGY CONSUMPTION: SINGLE FAMILY HOUSING. U.S. Government Printing Office, Washington, D.C. 1973.

1909. Hodgson, G.W., et al. DETECTION OF OIL LEAKS INTO WATER. Information Canada, Ottawa. 1973.

1910. Hogan, John D. "Resource Exploitation and Optimum Tax Policies: A Control Model Approach." In EXTRACTIVE RESOURCES AND TAXATION, Mason Gaffney (ed.). University of Wisconsin Press, Madison. 1967. (pp. 91-108).

1911. Hogan, Joseph W. AN ECONOMIC SIMULATION ANALYSIS OF THE USE OF DIGESTED SLUDGE TO RECLAIM STRIP-MINED LAND. Unpublished M.S. thesis. University of Illinois, Urbana. 1973.

1912. Hogarty, Thomas F. "The Geographic Scope of Energy Markets: Oil, Gas and Coal." Draft Report to the Energy Policy Project, Ford Foundation, New York. 1974.

1913. Holloman, J. Herbert, et al. "Energy R & D Policy Proposals." Draft Report to the Energy Policy Project. Ford Foundation, New York. July 1974.

1914. Holum, Ken. "Power Supply Using a National Grid." In INCREASING UNDERSTANDING OF PUBLIC PROBLEMS AND POLICIES, 1973. Fram Foundation, Chicago. 1973. (pp. 16-19).

1915. Houthakker, H.S. and Jorgenson, Dale W. "Energy Resources and Economic Growth." Draft Report to the Energy Policy Project. Ford Foundation, New York. September 1973.

1916. Howard, Herbert A. "A Measurement of the External Diseconomies Associated with Bituminous Coal Surface Mining, Eastern Kentucky, 1962-1967." NATURAL RESOURCES JOURNAL 11: 76-101. 1971.

1917. Hutchison, C.T. and Hellebust, J.A. OIL SPILLS AND VEGETATION AT NORMAL WELLS, N.W.T. Information Canada, Ottawa. 1974.

1918. Illinois Institute for Environmental Quality. STRIP MINE RECLAMATION IN ILLINOIS. Illinois Institute for Environmental Quality. 1973.

1919. Jones, F. "Depletion, Depreciation, and Coal Mining." TAXES 38: 31-42. 1960.

1920. Jovick, Robert L. CRITIQUE OF PROSPECTIVE COAL DEVELOPMENT IN EASTERN MONTANA. The Economic Development Association of Eastern Montana, Sidney. 1972.

1921. Keiffer, F.V. A BIBLIOGRAPHY OF SURFACE COAL MINING IN THE U.S. Forum Associates, Columbus, Ohio. 1972.

1922. Kent, Peter (ed.). ENERGY IN THE 1980'S. Royal Society, London. 1974.

1923. Kitch, E.W. "Regulation of the Field Market for Natural Gas by the Federal Power Commission." JOURNAL OF LAW AND ECONOMICS 11: 243-280. 1968.

1924. Krebs, Girard. "The Human Factor in Surface Mining." In REGULATION OF SURFACE MINING, PART II. U.S. Government Printing Office, Washington, D.C. 1973. (pp. 1080-1084).

1925. Kubo, A.S. and Rose, D.J. "Nuclear Waste Disposal." SCIENCE: 1205. December 21, 1973.

1926. Kuo, C.Y. A STUDY OF INCOME DISTRIBUTION IN THE MACKENZIE DISTRICT OF NORTHERN CANADA. Northern Economic Development Branch, Indian and Northern Affairs, Ottawa. 1973.

1927. Lantz, Herman R. PEOPLE OF COAL TOWN. Columbia University Press, New York. 1958.

1928. Large, David B. HIDDEN WASTE--POTENTIALS FOR ENERGY CONSERVATION. Conservation Foundation, Washington, D.C. 1974.

1929. Larkin, Peter A. "The Environmental Impact of Hydro-Power." In ENERGY AND THE ENVIRONMENT, Ian E. Efford and Barbara M. Smith (eds.). Institute of Resource Ecology, University of British Columbia, Vancouver. 1972. (pp. 162-175).

1930. Law, C.E., et al. RAILWAY TO THE ARCTIC: A STUDY OF THE OPERATIONS AND ECONOMIC FEASIBILITY OF A RAILWAY TO MOVE ARCTIC SLOPE OIL TO MARKET. Canadian Institute of Guided Ground Transport, Queen's University. 1972.

1931. Leistritz, F. Larry and Dalsted, Norman L. "North Dakota Coal Resources and Development Potential." NORTH DAKOTA FARM RESEARCH 31(6): 3-11. July-August 1973.

1932. Leistritz, F. Larry and Hertsgaard, Thor A. "Effects of Coal Development on Agriculture and Rural Communities in the Northern Great Plains." Paper Presented at the Annual Meeting of the American Agricultural Economics Association, Edmonton, Alberta. August 1973.

1933. Leistritz, F. Larry and Hertsgaard, Thor A. "Coal Development in North Dakota: Effects on Agriculture and Rural Communities." NORTH DAKOTA FARM RESEARCH 31(1): 3-9. September-October 1973.

1934. Leonard, Daniel. SOME ECONOMIC ASPECTS OF RECLAIMING STRIP-MINED LAND WITH DIGESTED SLUDGE. Unpublished M.S. thesis. University of Illinois. 1972.

1935. Lichtblau, John H. "The Outlook for Independent Domestic Refiners to the Early 1980's." Draft Report to the Energy Policy Project. Ford Foundation, New York. 1974.

1936. Little, Arthur D., Inc. COMPETITION IN THE NUCLEAR POWER SUPPLY INDUSTRY. U.S. Government Printing Office, Washington, D.C. 1968.

1937. Lockner, Allyn O. "The Economic Effect of the Severance Tax on the Decisions of the Mining Firm." NATURAL RESOURCES JOURNAL 4: 468-485. 1965.

1938. Lu, C.M. and Mathurin, D.C.E. POPULATION PROJECTIONS OF THE NORTHERN TERRITORIES TO 1981. Northern Policy and Program Planning Branch, Indian and Northern Affairs, Ottawa. 1973.

1939. Lubart, J.M. PSYCHODYNAMIC PROBLEMS OF ADAPTATION: MACKENZIE DELTA ESKIMOS. Northern Science Research Group, Indian and Northern Affairs, Ottawa. 1970.

1940. MacKay, D., Charles, M.E. and Phillips, C.R. CRUDE OIL SPILLS ON NORTHERN TERRAIN. Information Canada, Ottawa. 1973.

1941. McKelvey, Robert (ed.). A HAZARDOUS ENTERPRISE. University of Montana, Missoula. 1972.

1942. McLaren, Dale. IMPACT OF COAL MINING IN THE GREATER WABASH REGION. Greater Wabash Regional Planning Commission, Grayville, Illinois. 1973.

1943. McPhail, Robert L., et al. SOUTHWEST ENERGY STUDY, SUMMARY REPORT. U.S. Department of the Interior, Washington, D.C. 1972.

1944. Makhijani, A.B. and Lichtenberg, A.J. AN ASSESSMENT OF RESIDENTIAL ENERGY UTILIZATION IN THE U.S. Electronics Research Laboratory, University of California, Berkeley. 1973.

1945. Mancke, Richard B. THE FAILURE OF U.S. ENERGY POLICY. Columbia University Press, New York. 1974.

1946. Manders, P.M. AN EVALUATION OF THE ECONOMIC IMPACT OF THE MACKENZIE VALLEY GAS PIPELINE ON THE NORTHWEST TERRITORIES. Northern Policy and Program Planning Branch, Indian and Northern Affairs, Ottawa. 1973.

1947. Manning, Herbert C. "Mineral Rights Versus Surface Rights." NATURAL RESOURCES LAWYER 2: 329-346. 1969.

1948. Mansfield, Edwin. "Firm Size and Technological Change in the Petroleum and Bituminous Coal Industries." Draft Report to the Energy Policy Project. Ford Foundation, New York. 1973.

1949. Manula, Charles B. "Systems Simulation--A Gaming Model for Mine Management." MINING CONGRESS JOURNAL 49: 48-53. April 1963.

1950. Marine Engineers Beneficial Association. THE AMERICAN OIL INDUSTRY: A FAILURE OF ANTI-TRUST POLICY. Marine Engineers Beneficial Association, New York. 1973.

1951. Mead, Walter J. "Competition in the Energy Industry." Staff Paper. Energy Policy Project. Ford Foundation, New York. n.d.

1952. Mead, Walter J. "Natural Resource Disposal Policy--Oral Auction Versus Sealed Bids." NATURAL RESOURCES JOURNAL 7: 194-224. 1967.

1953. Meiners, Robert G. "Strip Mining Legislation." NATURAL RESOURCES JOURNAL 3: 442-469. 1964.

1954. Miller, E. Willard. "Strip Mining and Land Utilization in Western Pennsylvania." ECONOMIC GEOGRAPHY 28: 256-260. 1952.

1955. Moore, John R., et al. ECONOMICS OF THE PRIVATE AND SOCIAL COSTS OF APPALACHIAN COAL PRODUCTION: A PROGRESS REPORT. Appalachian Resources Project, University of Tennessee, Knoxville. n.d.

1956. Moore, Thomas Gale. "Economics of Scale and Firms Engaged in Oil and Coal Production." Draft Report to the Energy Policy Project. Ford Foundation, New York. 1973.

1957. Moore, Thomas Gale. "Potential Competition in Uranium Enriching." Draft Report to the Energy Policy Project. Ford Foundation, New York. 1973.

1958. Mooz, W.E. CALIFORNIA'S ELECTRICITY QUANDARY: 1. ESTIMATING FUTURE DEMAND. Rand Corporation, Santa Monica, California. 1972.

1959. Mooz, W.E. and Mow, C.C. A METHODOLOGY FOR PROJECTING THE ELECTRICAL ENERGY DEMAND OF THE COMMERCIAL SECTOR OF CALIFORNIA. Rand Corporation, Santa Monica, California. 1973.

1960. Mooz, W.E. and Mow, C.C. A METHODOLOGY FOR PROJECTING THE ELECTRICAL ENERGY DEMAND OF THE MANUFACTURING SECTOR OF CALIFORNIA. Rand Corporation, Santa Monica, California. 1973.

1961. Moyer, Reed. "Price-Output Behavior in the Coal Industry." Draft Report to the Energy Policy Project. Ford Foundation, New York. 1973.

1962. Mow, C.C., Mooz, W.E. and Anderson, S.K. A METHODOLOGY FOR PROJECTING THE ELECTRICAL ENERGY DEMAND OF THE RESIDENTIAL SECTOR IN CALIFORNIA. Rand Corporation, Santa Monica, California. 1973.

1963. Munn, Robert F. THE COAL INDUSTRY IN AMERICA: A BIBLIOGRAPHY AND GUIDE TO STUDIES. West Virginia University Library, Morgantown. 1965.

1964. Mutch, J.J. THE POTENTIAL FOR ENERGY CONSERVATION IN COMMERCIAL AIR TRANSPORT. Report Number R-1360-NSF. Rand Corporation, Santa Monica, California. 1973.

1965. Mutter, Douglas L. "The Regulation of Surface Mining: A Report on the Pending National Legislation." Federation of Rocky Mountain States, Denver, Colorado. 1973.

1966. Nathan, Robert T., Associates, Inc. THE POTENTIAL MARKET FOR FAR WESTERN COAL AND LIGNITE. U.S. Department of Commerce, Washington, D.C. 1965.

1967. Nathan, Robert T., Associates, Inc. THE POTENTIAL MARKET FOR MIDWESTERN AND ALASKAN COAL AND LIGNITE. U.S. Department of Commerce, Washington, D.C. 1966.

1968. National Academy of Sciences. REHABILITATION POTENTIAL OF WESTERN COAL LANDS. A Report to the Ford Foundation Energy Policy Project. Ballinger, Cambridge, Massachusetts. 1974.

1969. National Coal Association. SUMMARY OF STATE STRIP-MINING LAWS. National Coal Association, Washington, D.C. 1969.

1970. National Planning Association. DEMAND AND SUPPLY OF SCIENTIFIC AND TECHNICAL MANPOWER IN ENERGY-RELATED INDUSTRIES: UNITED STATES 1970-1985. National Planning Association. July 1974.

1971. National Petroleum Council. IMPACT OF NEW TECHNOLOGY ON THE U.S. PETROLEUM INDUSTRY, 1946-1965. National Petroleum Council, Washington, D.C. 1967.

1972. Natural Resources Defense Council. "Energy Decision Making in the Interior Department." Report to the Energy Policy Project. Ford Foundation, New York. 1973.

1973. Naysmith, J.K. CANADA NORTH: MAN AND THE LAND. Northern Economic Development Branch, Indian and Northern Affairs, Ottawa. 1971.

1974. Nellis, Lee. "What Does Energy Development Mean for Wyoming? A Community Study at Hanna, Wyoming." Office of Special Projects, University of Wyoming, Laramie. 1973.

1975. Nelson, David C. "A Study of the Demand for Electricity by Residential Consumers: Sample Markets in Nebraska." LAND ECONOMICS 41(1): 92-96. 1965.

1976. Nephew, E.A. SURFACE MINING AND LAND RECLAMATION IN GERMANY. Oak Ridge National Laboratory, Oak Ridge, Tennessee. 1972.

1977. O'Leary, John F. "Alternative Energy Sources and Environmental Conflict." In INCREASING UNDERSTANDING OF PUBLIC PROBLEMS AND POLICIES--1973. Farm Foundation, Chicago. 1973. (pp. 5-15).

1978. Ophuls, William. "The Scarcity Society: Farewell to the Free Lunch and to Freedom as an Infinite Resource." HARPERS 248: 47-52. April 1974.

1979. O'Riordan, T. "Public Opinion and Environmental Quality: A Reappraisal." ENVIRONMENT AND BEHAVIOR 3: 191-214. 1971.

1980. Over, J.A. (ed.). ENERGY CONSERVATION: WAYS AND MEANS. Publication Number 19. Future Shape of Technology Foundation, The Hague. 1974.

1981. Paddleford, D.F. "Analysis of Public Safety Risks Associated With Low Probability Nuclear Power Plant Accidents." Nuclear Energy Systems, Westinghouse Electric Corporation. September 1973.

1982. Palmer, J. SOCIAL ACCOUNTS FOR THE NORTH. Northern Policy and Program Planning Branch, Indian and Northern Affairs, Ottawa. 1973.

1983. Palmer, J. and St. Pierre, M. MONITORING SOCIO-ECONOMIC CHANGE: THE DESIGN AND TESTING OF A METHOD TO MONITOR AND ASSESS THE SOCIAL AND ECONOMIC EFFECTS OF PIPELINE DEVELOPMENT ON COMMUNITIES ON THE ROUTE. Report Number 74-7. Northern Policy and Program Planning Branch, Indian and Northern Affairs. Information Canada, Ottawa. 1974.

1984. Parker, Albert. "World Energy Prospects: An Appraisal." FUEL 49: 289-308. 1970.

1985. Parsons, G. ARCTIC SUBURB: A LOOK AT THE NORTH'S NEWCOMERS. Northern Science Research Group, Indian and Northern Affairs, Ottawa. 1970.

1986. Payne, Christopher (ed.). FUEL AND THE ENVIRONMENT. Ann Arbor Science Publishers, Ann Arbor, Michigan. 1974.

1987. Persse, Franklin H. STRIP-MINING TECHNIQUES TO MINIMIZE ENVIRONMENTAL DAMAGE IN THE UPPER MISSOURI RIVER BASIN STATES. Preliminary Report Number 192. Bureau of Mines, U.S. Department of the Interior, Washington, D.C. 1973.

1988. Persse, Franklin H. and Toland, Joseph E. IMPACT OF ENVIRONMENTAL POLICIES ON USE OF UPPER MISSOURI RIVER BASIN COAL, LIGNITE, AND WATER. Preliminary Report Number 188. Bureau of Mines, U.S. Department of the Interior, Washington, D.C. 1972.

1989. Peters, W.C. (ed.). MINING AND ECOLOGY IN THE ARID ENVIRONMENT. Proceedings of a Symposium, College of Mines, University of Arizona, Tucson. 1970.

1990. Petruschell, R.L. and Salter, R.G. ELECTRICITY GENERATING COST MODEL FOR COMPARISON OF CALIFORNIA POWER PLANT SITE ALTERNATIVES. R-1087-RF/CSA. Rand Corporation, Santa Monica, California. 1973.

1991. Pickard, Claude Eugene. THE WESTERN KENTUCKY COAL FIELDS: THE INFLUENCE OF COAL MINING ON SETTLEMENT PATTERNS, FORMS, AND FUNCTIONS. Unpublished Ph.D. dissertation, University of Nebraska, Lincoln. 1969.

1992. Pimentel, David. "Energy Crisis and Crop Production." In ENERGY AND AGRICULTURE: RESEARCH IMPLICATIONS, Loyd Fischer and Arlo Biere (eds.). North Central Research Strategy Committee on Natural Resource Development. October 1973. (pp. 65-79).

1993. Pletsch, D.C., et al. "The Trans-Alaska Pipeline." In ARCTIC OIL AND GAS: PROBLEMS AND POSSIBILITIES. Fifth International Congress. Fondation Francaise d'Etudes Nordiques. 1973.

1994. Power, Thomas M. "Federal Regulation of Strip Mining: Doubtful Protection." MONTANA BUSINESS QUARTERLY 11(3): 21-25. 1973.

1995. Putman, Palmer Cosslett. POWER FROM THE WIND. Van Nostrand, Reinhold, New York. 1974.

1996. Radforth, J.R. ANALYSIS OF DISTURBANCE EFFECTS OF OPERATIONS OF OFF-ROAD VEHICLES ON TUNDRA. Public Information, Indian and Northern Affairs, Ottawa. 1972.

1997. Reardon, W.A. and Wilfert, G.L. AN ANALYSIS OF U.S. ENERGY CONSUMPTION FROM 1947 to 1958. Battelle Memorial Institute, Richland, Washington. 1970.

1998. Reeve, A.J. "Some Sociological Implications of Pipeline Construction." In PROCEEDINGS OF THE CANADIAN NORTHERN PIPELINE RESEARCH CONFERENCE, 2-4 February, 1972, R.F. Leggett, R.F. Macfarlane, and I.C. Macfarlane (eds.). Technical Memorandum 104. National Research Council of Canada, Ottawa. 1972.

1999. Renkey, Leslie E. "Local Zoning of Strip Mining." KENTUCKY LAW JOURNAL 57: 738-755. 1969.

2000. Reitze, Arnold W., Jr. "Old King Coal and the Merry Rapists of Appalachia." CASE WESTERN RESERVE LAW REVIEW 22: 650-737. 1971.

2001. Reitze, Arnold W., Jr. ENVIRONMENTAL LAW. North American International, Washington, D.C. 1972.

2002. Reitze, Arnold W., Jr. ENVIRONMENTAL PLANNING: LAW OF LAND AND RESOURCES. North American International, Washington, D.C. 1974.

2003. Resources for the Future. "Toward Self-Sufficiency in Energy Supply." Draft Report to the Energy Policy Project. Ford Foundation, New York. September 1973.

2004. Roberts, Keith (ed.). TOWARDS AN ENERGY POLICY. Sierra Club, San Francisco. 1973.

2005. Roseberry, John L. and Klimstra, W.D. "Recreational Activities on Illinois Strip-Mined Lands." JOURNAL OF SOIL AND WATER CONSERVATION 19: 107-109. 1964.

2006. Sagan, Leonard A. (ed.). HUMAN AND ECOLOGIC EFFECTS OF NUCLEAR POWER PLANTS. Charles C Thomas, Springfield, Illinois. 1974.

2007. Salamon, Lester M. and Siegfried, John J. "The Relationship Between Economic Structure and Political Power: The Energy Industry." Draft Report to the Energy Policy Project. Ford Foundation, New York. 1973.

2008. Schwartz, Donald. "Lignite in the Economy." In SYMPOSIUM ON THE GREAT PLAINS OF NORTH AMERICA, Carle C. Zimmerman and Seth Russell (eds.). The North Dakota Institute for Regional Studies, Fargo. 1967. (pp. 125-129).

2009. Scott, M. THE SOCIO-ECONOMIC IMPACT OF THE POINTED MOUNTAIN GASFIELD. Northern Policy and Planning A.C.N.D. Division, Indian and Northern Affairs, Ottawa. 1973.

2010. Seitz, Wesley D. "An Analysis of Strip Mining and Local Taxation Practices." ILLINOIS AGRICULTURAL ECONOMICS 12(1): 23-30. 1972.

2011. Seitz, Wesley D. "An Economic Evaluation of the Application of Sewage Sludge on Strip-Mined Land." Paper Presented at the Annual Meeting of the American Agricultural Economics Association, Edmonton, Alberta. August 1973.

2012. Shaw, Elmer W. (Comp.). A REVIEW OF THE ENERGY RESOURCES OF THE PUBLIC LANDS BASED ON STUDIES SPONSORED BY THE PUBLIC LAND LAW REVIEW COMMISSION. Prepared for the Committee on Interior and Insular Affairs. U.S. Senate. U.S. Government Printing Office, Washington, D.C. 1971.

2013. Shrum, Gordon M. "Meeting British Columbia's Energy Requirements." In ENERGY AND THE ENVIRONMENT, Ian E. Efford and Barbara M. Smith (eds.). Institute of Resource Ecology, University of British Columbia, Vancouver. 1972. (pp. 11-31).

2014. Siehl, George H. THE ISSUES RELATED TO SURFACE MINING. U.S. Government Printing Office, Washington, D.C. 1971.

2015. Siehl, George H. LEGISLATIVE PROPOSALS CONCERNING SURFACE MINING OF COAL. U.S. Government Printing Office, Washington, D.C. 1971.

2016. Smith, Courtland L. "Self-Interest Groups and Human Emotion as Adaptive Mechanisms." In WATER AND COMMUNITY DEVELOPMENT: SOCIAL AND ECONOMIC PERSPECTIVES, Donald R. Field, James C. Barron and Burl F. Long (eds.). Ann Arbor Science Publishers, Inc., Ann Arbor, Michigan. 1974. (pp. 151-168).

2017. Smith, D.G. OCCUPATIONAL PREFERENCES OF NORTHERN STUDENTS. Northern Science Research Group, Indian and Northern Affairs, Ottawa. 1974.

2018. Smith, V. "The Economics of Production from Natural Resources." AMERICAN ECONOMIC REVIEW 58: 409-431. June 1968.

2019. Spaulding, Willard M., Jr. and Ogden, Ronald D. EFFECTS OF SURFACE MINING ON THE FISH AND WILDLIFE RESOURCES OF THE UNITED STATES. Resource Publication 68. Bureau of Sport Fisheries and Wildlife, Fish and Wildlife Service, U.S. Department of the Interior. U.S. Government Printing Office, Washington, D.C. 1968.

2020. Spore, Robert L. "Evaluation of Alternative Land Use: Coal Surface Mining vs. Natural Environment Preservation." In REGULATION OF SURFACE MINING, PART II. U.S. Government Printing Office, Washington, D.C. 1973. (pp. 1287-1294).

2021. Stager, J.K. OLD CROW, YUKON TERRITORY AND THE PROPOSED ARCTIC GAS PIPELINE. Social Research Division, Indian and Northern Affairs, Ottawa. 1974.

2022. Steinhart, John S. and Steinhart, Carol E. "Energy Use in the U.S. Food System." In ENERGY AND AGRICULTURE: RESEARCH IMPLICATIONS, Loyd Fischer and Arlo Biere (eds.). North Central Research Strategy Committee on Natural Resource Development. October 1973. (pp. 41-64).

2023. Stevenson, D.S. PROBLEMS OF ESKIMO RELOCATION FOR INDUSTRIAL EMPLOYMENT. Northern Science Research Group, Indian and Northern Affairs, Ottawa. 1968.

2024. Stewart, Charles L. "Strategy in Protecting the Public's Interest in Land: With Special Reference to Strip Mining." THE JOURNAL OF LAND AND PUBLIC UTILITY ECONOMICS 15: 312-316. 1939.

2025. Stewart, Earl E. and Stewart, Robert E. A MULTIPLE LAND USE STUDY FOR A NINE-COUNTY AREA OF SOUTHWESTERN NORTH DAKOTA. Little Missouri Grasslands Study, North Dakota State University, Fargo. 1974.

2026. Stewart, Robert E., Jr. (ed.). CONFERENCE ON PROPERTY AND MINERAL RIGHTS. Interim Report Number 2. Little Missouri Grasslands Study, North Dakota State University, Fargo. 1973.

2027. Stoltžfus, Victor. "Social Impacts of Energy Rationing Alternatives." In ENERGY AND AGRICULTURE: RESEARCH IMPLICATIONS, Loyd Fischer and Arlo Biere (eds.). North Central Research Strategy Committee on Natural Resource Development. October 1973. (pp. 80-94).

2028. Stoner, Carol Hupping (ed.). PRODUCING YOUR OWN POWER: HOW TO MAKE NATURE'S ENERGY SOURCES WORK FOR YOU. Rodale Press, Emmaus, Pennsylvania. 1974.

2029. Stork, Karen E. THE ROLE OF WATER IN THE ENERGY CRISIS. Nebraska Water Resources Research Institute, Lincoln, Nebraska. 1973.

2030. Sullivan, R.W., et al. A BRIEF OVERVIEW OF ENERGY REQUIREMENTS FOR THE DEPARTMENT OF DEFENSE. Battelle Columbus Laboratories, Columbus, Ohio. 1972.

2031. Swanson, Earl R. ECONOMIC IMPACT OF REGULATIONS TO LIMIT USE OF NITROGEN FERTILIZER. Special Publication Number 26. College of Agriculture, University of Illinois. May 1972.

2032. Thibault, E. REGIONAL SOCIO-ECONOMIC OVERVIEW STUDY: YUKON. Government of the Yukon. 1974.

2033. Thirring, H. ENERGY FOR MAN. Harper Torchbooks, New York. 1958.

2034. Thrush, Paul W., et al. A DICTIONARY OF MINING, MINERALS AND RELATED TERMS. Bureau of Mines, U.S. Department of the Interior, Washington, D.C. 1968.

2035. Townsend, Stuart and VanLanen, James. AN ECONOMIC ANALYSIS OF ALTERNATIVE TECHNOLOGIES FOR COOLING THERMAL-ELECTRIC GENERATING PLANTS IN THE FORT UNION COAL REGION. Research Report Number 48. Montana Agricultural Experiment Station, Bozeman. 1974.

2036. Travacon Research. ECONOMIC STUDY OF TRANSPORTATION IN THE MACKENZIE RIVER VALLEY. Prepared for the Transportation Development Agency, Transport Canada, Ottawa. 1972.

2037. Turk, Jonathan, et al. ECOSYSTEMS, ENERGY, POPULATION. W.B. Saunders, Philadelphia. (Forthcoming February 1975).

2038. Tybout, Richard A. "Electric Power Rates and the Environment." In THE ELECTRIC POWER INDUSTRY AND THE ENVIRONMENT, Sierra Club. 1973. (pp. 527-580).

2039. Udall, Stewart, Conconi, Charles and Osterhout, David. THE ENERGY BALLOON. McGraw-Hill, New York. 1974.

2040. U.S. Atomic Energy Commission. THE SAFETY OF NUCLEAR POWER REACTORS AND RELATED FACILITIES. U.S. Government Printing Office, Washington, D.C. 1973.

2041. U.S. Comptroller General. IMPROVEMENTS NEEDED IN ADMINISTRATION OF FEDERAL COAL LEASING PROGRAM. U.S. General Accounting Office, Washington, D.C. 1972.

2042. U.S. Congress. ENVIRONMENTAL EFFECTS OF PRODUCING ELECTRIC POWER. Hearings before the Joint Atomic Energy Committee. 91st Congress. U.S. Government Printing Office, Washington, D.C. 1970.

2043. U.S. Congress. PRELICENSING ANTITRUST REVIEW OF NUCLEAR POWERPLANTS. Hearings before the Joint Atomic Energy Committee. 91st Congress. U.S. Government Printing Office, Washington, D.C. 1970.

2044. U.S. Congress. FUTURE STRUCTURE OF THE URANIUM ENRICHMENT INDUSTRY. Hearings before the Joint Atomic Energy Committee. 93rd Congress. U.S. Government Printing Office, Washington, D.C. 1973.

2045. U.S. Congress. HIGHLIGHTS OF ENERGY LEGISLATION IN THE 93RD CONGRESS, 1ST SESSION. U.S. Government Printing Office, Washington, D.C. 1974.

2046. U.S. Congress. NATURAL GAS REGULATION AND THE TRANS-ALASKA PIPELINE. Hearings before the Joint Economic Committee. 92nd Congress. U.S. Government Printing Office, Washington, D.C. 1972.

2047. U.S. Congress. ECONOMIC IMPACT OF PETROLEUM SHORTAGES. Hearings before the Subcommittee on International Economics of the Joint Economic Committee. 93rd Congress. U.S. Government Printing Office, Washington, D.C. 1974.

2048. U.S. Congress. House. OIL IMPORT CONTROLS. Hearings before the Subcommittee on Mines and Mining of the Committee on Interior and Insular Affairs. 91st Congress. U.S. Government Printing Office, Washington, D.C. 1970.

2049. U.S. Congress. House. POWERPLANT SITING AND ENVIRONMENTAL PROTECTION. Hearings before the Subcommittee on Communications and Power of the Committee on Interstate and Foreign Commerce. 92nd Congress. U.S. Government Printing Office, Washington, D.C. 1971.

2050. U.S. Congress. House. FOREIGN POLICY IMPLICATIONS OF THE ENERGY CRISIS. Hearings before the Subcommittee on Foreign Economic Policy of the Committee on Foreign Affairs. 92nd Congress. U.S. Government Printing Office, Washington, D.C. 1972.

2051. U.S. Congress. House. FUEL AND ENERGY RESOURCES, 1972. Hearings before the Committee on Interior and Insular Affairs. U.S. Government Printing Office, Washington, D.C. 1972.

2052. U.S. Congress. House. ADVERSE EFFECTS OF COAL MINING ON FEDERAL RESERVOIR PROJECTS. Hearings before a Subcommittee of the Committee on Government Operations. 93rd Congress. U.S. Government Printing Office, Washington, D.C. 1973.

2053. U.S. Congress. House. ENERGY EMERGENCY ACT. Hearings before the Committee on Interstate and Foreign Commerce. 93rd Congress. U.S. Government Printing Office, Washington, D.C. 1973.

2054. U.S. Congress. House. ENERGY CRISIS AND ITS EFFECT ON AGRICULTURE. Hearings before the Committee on Agriculture. 93rd Congress. U.S. Government Printing Office, Washington, D.C. 1973.

2055. U.S. Congress. House. EPA POLLUTION REGULATIONS AND FUEL SHORTAGE: THE IMPACT ON MASS TRANSIT. Hearing before the Subcommittee on Urban Mass Transportation of the Committee on Banking and Currency. 93rd Congress. U.S. Government Printing Office, Washington, D.C. 1973.

2056. U.S. Congress. House. MANDATORY FUELS ALLOCATION. Hearings before the Committee on Interstate and Foreign Commerce. 93rd Congress. U.S. Government Printing Office, Washington, D.C. 1973.

2057. U.S. Congress. House. OIL AND NATURAL GAS PIPELINE RIGHTS-OF-WAY. Hearings before the Subcommittee on Public Lands of the Committee on Interior and Insular Affairs. 93rd Congress. U.S. Government Printing Office, Washington, D.C. 1973.

2058. U.S. Congress. House. PHASE IV OIL REGULATIONS AND PETROLEUM MARKETING PROBLEMS. Hearings before the Subcommittee on Activities of Regulatory Agencies of the Permanent Select Committee on Small Business. 93rd Congress. U.S. Government Printing Office, Washington, D.C. 1973.

2059. U.S. Congress. House. REGULATION OF SURFACE MINING. Hearings before the Subcommittee on the Environment of the Subcommittee on Mines and Mining of the Committee on Interior and Insular Affairs. 93rd Congress. U.S. Government Printing Office, Washington, D.C. 1973.

2060. U.S. Congress. House. Committee on Interior and Insular Affairs. SURFACE MINING CONTROL AND RECLAMATION ACT OF 1974. 93rd Congress. U.S. Government Printing Office, Washington, D.C. 1974.

2061. U.S. Congress. House. Committee on Science and Astronautics. SOLAR ENERGY RESEARCH: A MULTIDISCIPLINARY APPROACH. 92nd Congress. U.S. Government Printing Office, Washington, D.C. 1972.

2062. U.S. Congress. Senate. POWER POLICY. Hearings before the Committee on the Judiciary. 83rd Congress. U.S. Government Printing Office, Washington, D.C. 1954.

2063. U.S. Congress. Senate. AMENDMENTS TO THE NATURAL GAS ACT. Hearings before the Committee on Interstate and Foreign Commerce. 84th Congress. U.S. Government Printing Office, Washington, D.C. 1955.

2064. U.S. Congress. ENERGY RESOURCES AND TECHNOLOGY. Hearings before the Subcommittee on Automation and Energy Resources of the Joint Economic Committee. 86th Congress. U.S. Government Printing Office, Washington, D.C. 1959.

2065. U.S. Congress. Senate. OIL IMPORT ALLOCATIONS. Hearings before the Select Committee on Small Business. 88th Congress. U.S. Government Printing Office, Washington, D.C. 1964.

2066. U.S. Congress. Senate. TRANS-ALASKA PIPELINE. Hearings before the Committee on Interior and Insular Affairs. 91st Congress. U.S. Government Printing Office, Washington, D.C. 1969.

2067. U.S. Congress. Senate. NATIONAL ENVIRONMENTAL POLICY ACT RELATIVE TO HIGHWAYS. Hearings before the Subcommittee on Roads of the Committee on Public Works. 91st Congress. U.S. Government Printing Office, Washington, D.C. 1970.

2068. U.S. Congress. Senate. OIL SHALE DEVELOPMENT. Hearings before the Subcommittee on Minerals, Materials and Fuels of the Committee on Interior and Insular Affairs. 91st Congress. U.S. Government Printing Office, Washington, D.C. 1970.

2069. U.S. Congress. Senate. SANTA BARBARA OIL POLLUTION. Hearings before the Subcommittee on Minerals, Materials and Fuels of the Committee on Interior and Insular Affairs. 91st Congress. U.S. Government Printing Office, Washington, D.C. 1970.

2070. U.S. Congress. Senate. USE OF HIGHWAY FUNDS FOR PUBLIC TRANSPORTATION. Hearings before the Subcommittee on Roads of the Committee on Public Works. 91st Congress. U.S. Government Printing Office, Washington, D.C. 1970.

2071. U.S. Congress. Senate. COMBINATION UTILITY COMPANIES. Hearings before the Subcommittee on Antitrust and Monopoly of the Committee on the Judiciary. 92nd Congress. U.S. Government Printing Office, Washington, D.C. 1971.

2072. U.S. Congress. Senate. MINING ACTIVITIES IN THE CUSTER AND CALLATIN NATIONAL FORESTS IN MONTANA. Hearings before the Subcommittee on Minerals, Materials and Fuels of the Committee on Interior and Insular Affairs. 92nd Congress. U.S. Government Printing Office, Washington, D.C. 1971.

2073. U.S. Congress. Senate. MINING AND MINERALS POLICY ACT OF 1970. Hearings before the Subcommittee on Minerals, Materials and Fuels of the Committee on Interior and Insular Affairs. 92nd Congress. U.S. Government Printing Office, Washington, D.C. 1971.

2074. U.S. Congress. Senate. NATURAL GAS SUPPLY FOR THE PACIFIC NORTHWEST. Hearings before the Committee on Commerce. 92nd Congress. U.S. Government Printing Office, Washington, D.C. 1971.

2075. U.S. Congress. Senate. SURFACE MINING. Hearings before the Subcommittee on Minerals, Materials and Fuels of the Committee on Interior and Insular Affairs. 92nd Congress. U.S. Government Printing Office, Washington, D.C. 1971.

2076. U.S. Congress. Senate. ALTERNATIVES TO THE GASOLINE-POWERED INTERNAL COMBUSTION ENGINE. Hearings before the Panel on Environmental Science and Technology of the Subcommittee on Air and Water Pollution of the Committee on Public Works. 92nd Congress. U.S. Government Printing Office, Washington, D.C. 1972.

2077. U.S. Congress. Senate. COUNCIL ON ENERGY POLICY. Joint hearings before the Committee on Commerce and the Committee on Interior and Insular Affairs. 92nd Congress. U.S. Government Printing Office, Washington, D.C. 1972.

2078. U.S. Congress. Senate. ENERGY RESEARCH AND DEVELOPMENT. Hearings before the Committee on Commerce. 92nd Congress. U.S. Government Printing Office, Washington, D.C. 1972.

2079. U.S. Congress. Senate. ENERGY RESEARCH AND DEVELOPMENT POLICY ACT. Hearings before the Committee on Interior and Insular Affairs. 93rd Congress. U.S. Government Printing Office, Washington, D.C. 1973.

2080. U.S. Congress. Senate. THE IMPACT OF AUTO EMISSION STANDARDS. Report by the staff of the Subcommittee on Air and Water Pollution to the Committee on Public Works. 93rd Congress. U.S. Government Printing Office, Washington, D.C. 1973.

2081. U.S. Congress. Senate. IMPACT OF THE FUEL SHORTAGE ON AGRICULTURE. Hearings before the Subcommittee on Agricultural Research and General Legislation of the Committee on Agriculture and Forestry. 93rd Congress. U.S. Government Printing Office, Washington, D.C. 1973.

2082. U.S. Congress. Senate. INTERNATIONAL COMPENSATION FUND FOR OIL POLLUTION DAMAGE. Hearings before the Subcommittee on Oceans and International Environment of the Committee on Foreign Relations. 92nd Congress. U.S. Government Printing Office, Washington, D.C. 1973.

2083. U.S. Congress. Senate. PETROLEUM PRODUCT SHORTAGES. Hearings before the Committee on Banking, Housing and Urban Affairs. 93rd Congress. U.S. Government Printing Office, Washington, D.C. 1973.

2084. U.S. Congress. Senate. REGULATION OF SURFACE MINING OPERATIONS. Hearings before the Committee on Interior and Insular Affairs. 93rd Congress. U.S. Government Printing Office, Washington, D.C. 1973.

2085. U.S. Congress. Senate. COMPENSATION FOR UNEMPLOYMENT RELATED TO THE ENERGY CRISIS. Hearings before the Committee on Finance. 93rd Congress. U.S. Government Printing Office, Washington, D.C. 1974.

2086. U.S. Congress. Senate. ENERGY AND ENVIRONMENT OBJECTIVES. Hearings before the Subcommittee on Environment of the Committee on Commerce. 93rd Congress. U.S. Government Printing Office, Washington, D.C. 1974.

2087. U.S. Congress. Senate. OIL COMPANY PROFITABILITY. Hearings before the Committee on Finance. 93rd Congress. U.S. Government Printing Office, Washington, D.C. 1974.

2088. U.S. Congress. Senate. TRANSPORTATION AND THE NEW ENERGY POLICIES. Hearings before the Subcommittee on Transportation of the Committee on Public Works. 93rd Congress. U.S. Government Printing Office, Washington, D.C. 1974.

2089. U.S. Congress. Senate. Committee on Interior and Insular Affairs. FEDERAL RESOURCES (FUNDING AND PERSONNEL) IN ENERGY RELATED ACTIVITIES, FISCAL YEARS 1972 AND 1973. 92nd Congress. U.S. Government Printing Office, Washington, D.C. 1972.

2090. U.S. Congress. Senate. Committee on Interior and Insular Affairs. A REVIEW OF ENERGY POLICY ACTIVITIES OF THE 92ND CONGRESS, 1ST SESSION. 92nd Congress. U.S. Government Printing Office, Washington, D.C. 1972.

2091. U.S. Congress. Senate. Committee on Interior and Insular Affairs. SURFACE MINING RECLAMATION ACT OF 1973. 93rd Congress. U.S. Government Printing Office, Washington, D.C. 1973.

2092. U.S. Congress. Senate. Committee on Interior and Insular Affairs. SURVEY OF THE ENERGY CONSERVATION AND DEVELOPMENT RECOMMENDATIONS CONTAINED IN THE FINAL REPORT OF THE NATIONAL COMMISSION ON MATERIALS POLICY. 93rd Congress. U.S. Government Printing Office, Washington, D.C. 1973.

2093. U.S. Congressional Research Service: Environmental Policy Division. CONGRESS AND THE NATION'S ENVIRONMENT. U.S. Government Printing Office, Washington, D.C. 1973.

2094. U.S. Council on Environmental Quality. COAL SURFACE MINING AND RECLAMATION: AN ENVIRONMENTAL AND ECONOMIC ASSESSMENT OF ALTERNATIVES. U.S. Government Printing Office, Washington, D.C. 1973.

2095. U.S. Council on Environmental Quality. ENERGY AND THE ENVIRONMENT: ELECTRIC POWER. U.S. Government Printing Office, Washington, D.C. 1973.

2096. U.S. Environmental Protection Agency, Rocky Mountain-Prairie Region. A PRELIMINARY BIBLIOGRAPHY OF PUBLICATIONS CONCERNING REHABILITATION OF LANDS DISTURBED BY MINING AND ASSOCIATED ACTIVITIES FOR THE NORTHERN GREAT PLAINS RESOURCES PROGRAM. U.S. Environmental Protection Agency, Denver, Colorado. 1973.

2097. U.S. Federal Energy Administration. PROJECT INDEPENDENCE REPORT. U.S. Government Printing Office, Washington, D.C. 1974.

2098. U.S. Federal Trade Commission. THE INTERNATIONAL PETROLEUM CARTEL. U.S. Government Printing Office, Washington, D.C. 1952.

2099. U.S. Government. ENERGY CONSERVATION HANDBOOK FOR LIGHT INDUSTRIES AND COMMERCIAL BUILDINGS. U.S. Government Printing Office, Washington, D.C. 1974.

2100. U.S. Government. ENERGY POLICY PAPERS. U.S. Government Printing Office, Washington, D.C. 1974.

2101. U.S. Government. THE NATION'S ENERGY FUTURE. U.S. Government Printing Office, Washington, D.C. 1974.

2102. U.S. Government. THE PROSPECTS FOR GASOLINE AVAILABILITY: 1974. U.S. Government Printing Office, Washington, D.C. 1974.

2103. U.S. National Aeronautics and Space Administration. ENERGY: A SPECIAL BIBLIOGRAPHY WITH INDEXES. National Technical Information Service, Springfield, Virginia. 1974.

2104. University of Maryland. School of Law. LEGAL PROBLEMS OF COAL MINE RECLAMATION. For the Environmental Protection Agency. U.S. Government Printing Office, Washington, D.C. 1972.

2105. Usher, P.J. THE BANKSLANDERS: ECONOMY AND ECOLOGY OF A FRONTIER TRAPPING COMMUNITY. Northern Science Research Group, Indian and Northern Affairs, Ottawa. 1972.

2106. Usher, P.J. and Beakhust, G. LAND REGULATION IN THE CANADIAN NORTH. Canadian Arctic Resources Committee, Ottawa. 1973.

2107. Van Lanen, James L., McConnen, R.J. and Thompson, Layton S. "Estimated Impact on Tax Revenues and Coal Production of Severance Taxes on Montana Coal." Department of Agricultural Economics and Economics, University of Montana, Bozeman. 1973.

2108. Veatch, Henry B. "What's Energy to Ethics or Ethics to Energy?" Paper Presented at the Energy Conference, Halifax, Nova Scotia. August 1974.

2109. J.M. Viladas Company. THE AMERICAN PEOPLE AND THEIR ENVIRONMENT. U.S. Environmental Protection Agency, Washington, D.C. 1974.

2110. J.M. Viladas Company. IMPACT OF THE FUEL SHORTAGE ON PUBLIC ATTITUDES TOWARD ENVIRONMENTAL PROTECTION. Report Submitted to the Environmental Protection Agency. U.S. Environmental Protection Agency, Washington, D.C. 1974.

2111. Villiers, D. THE CENTRAL MACKENZIE: AN AREA ECONOMIC SURVEY. Northern Administration Branch, Indian and Northern Affairs, Ottawa. 1967.

2112. Wali, Mohan K. (ed.). SOME ENVIRONMENTAL ASPECTS OF STRIP MINING IN NORTH DAKOTA. Educational Series 5. North Dakota Geological Survey, Grand Forks. 1974.

2113. Walker, Melvin E., Jr. AN ECONOMIC EVALUATION OF THE IMPACT OF COMMERCIAL NITROGEN CONTROL AT THE FARM LEVEL. Unpublished Ph.D. dissertation. Department of Agricultural Economics, University of Illinois. 1973.

2114. Walter, George H. "Agriculture and Strip Coal Mining." AGRICULTURAL ECONOMICS RESEARCH 1: 24-29. 1949.

2115. Weingart, Jerome, Shoen, Richard, et al. "Institutional Problems of the Application of New Community Energy System Technologies." Caltech Environmental Quality Laboratory Draft Report to the Energy Policy Project. Ford Foundation, New York. November 1973.

2116. White, David C. "The Energy-Environment-Economic Triangle." TECHNOLOGY REVIEW 25(12): 11-19. December 1973.

2117. White, William C. "The Impact of Environmental Control on the Fertilizer Industry." AGRICULTURE BUSINESS FORUM PROCEEDINGS. August 1972.

2118. Wicks, Gary, et al. "A Look at Coal-Related Legislation." MONTANA BUSINESS QUARTERLY 11(3): 5-15. 1973.

2119. Williams, J. Richard. SOLAR ENERGY: TECHNOLOGY AND APPLICATIONS. Ann Arbor Science Publishers, Ann Arbor, Michigan. 1974.

2120. Wilson, Richard and Jones, William J. ENERGY, ECOLOGY AND THE ENVIRONMENT. Academic Press, New York. 1974.

2121. Winger, John G., et al. OUTLOOK FOR ENERGY IN THE UNITED STATES TO 1985. Energy Economics Division, Chase Manhattan Bank, New York. 1972.

2122. Wolforth, J.R. THE MACKENZIE DELTA: ITS ECONOMIC BASE AND DEVELOPMENT. Northern Coordination and Research Centre, Indian and Northern Affairs, Ottawa. 1966.

2123. Wolforth, J.R. THE EVOLUTION AND ECONOMY OF THE DELTA COMMUNITY. Northern Science Research Group, Indian and Northern Affairs, Ottawa. 1971.

2124. Ziegler, George. "Coal Mining in Ohio and Its Effects on Environmental Health." Ohio Department of Health, Columbus. 1965.

SUBJECT INDEX CATEGORIES

SUBJECT INDEX CATEGORIES

Aesthetic, humanistic, literary, religious, philosophic
Africa
Agriculture, food, ranching, rural
Air
Alaska/pipeline
Alternative sources, general
Anthropology
Appalachia
Attitude, behavior, opinions, motives, values, perceptions, cognitions, knowledge, psychology
Automobile
Built environment, urban environment
Canada
Coal/shale
Communication, media
Conflicts, controversy, competition
Conservation
Economics, business, industry, work, occupations
Education
Electricity
Energy crisis
Energy, general
Energy industry associations
England
Environment and energy
Equity problems
Europe, Eastern (including Russia)
Europe, Western
Forests
Fossil fuels, general
General ecology, social ecology, human ecology, eco-systems

Geography, regional studies
Geothermal
Government, public agencies
Growth and energy
History
Hydroelectric
Import/export
International, interstate, intergovernmental, interagency
Land
Law, property rights
Management, policy, decision making, planning, development
Mass/public transit
Methodology, evaluation, measurement, indicators, systems
  analysis, cost-benefits techniques, projections, monitoring,
  control, standards, performance criteria, theory, concepts,
  research and development
Middle East/Arabs
Military, war, security
Natural or bottle gas
Natural resources, general
North Sea
Nuclear
Offshore/ocean sources
Oilspills
Petroleum
Politics, political science
Pollution, general
Population, demography, migration, crowding
Quality of life, affluence, living standards, general
  environmental quality, life-style
Railroads
Readers, conference proceedings, special issues, textbooks,
  bibliographies
Reclamation
Recreation, leisure, parks, wilderness, wildlife, nature

Safety, health

Science, technology

Siting

Sociology, social organization, institutions, culture, society

Solar

Strip mining

Supply/demand/price

Tankers/tanker spills

Taxation

Utilities

Voluntary action, voluntary organizations, citizen participation, social movements

Waste

Water

Western/Southwestern U.S.

Thermal energy, thermal pollution

Transportation, general

SUBJECT INDEX

SUBJECT INDEX

Aesthetic, humanistic, literary, religious, philosophic, 65, 238, 239, 286, 303, 401, 406, 416, 427, 740, 764, 791, 870, 888, 949, 1137, 1158, 1269, 1384, 1492, 1717, 1818, 1827, 1878, 1927, 2006, 2108

Africa, 587, 1158, 1341, 1689

Agriculture, food, ranching, rural, 2, 93, 227, 412, 416, 450, 523, 524, 644, 726, 916, 1057, 1089, 1103, 1139, 1147, 1148, 1149, 1175, 1211, 1339, 1341, 1405, 1415, 1460, 1522, 1639, 1640, 1720, 1797, 1818, 1852, 1863, 1893, 1896, 1907, 1911, 1931, 1932, 1933, 1992, 2010, 2011, 2022, 2027, 2031, 2054, 2081, 2113, 2114, 2117

Air, 268, 317, 324, 383, 553, 649, 723, 724, 790, 925, 1038, 1218, 1239, 1539, 1644, 1655, 2080

Alaska/pipeline, 19, 49, 260, 261, 311, 315, 316, 634, 798, 986, 1205, 1215, 1248, 1268, 1317, 1725, 1810, 1855, 1900, 1940, 1967, 1983, 1993, 1998, 2046, 2066

Alternative sources, general, 57, 92, 303, 330, 427, 501, 561, 575, 595, 600, 601, 607, 631, 663, 670, 803, 810, 811, 854, 921, 957, 1032, 1193, 1222, 1276, 1302, 1327, 1381, 1451, 1479, 1539, 1559, 1653, 1757, 1823, 1892, 1977, 2028, 2076

Anthropology, 8, 130, 318, 363, 395, 616, 743, 805, 1210, 1211, 1222, 1275, 1336, 1461, 1726, 1737, 1738, 1822, 1939, 1973, 2105

Appalachia, 63, 238, 239, 346, 1157, 1798, 1803, 1844, 1905, 1916, 1927, 1955, 1991, 2000

Attitude, behavior, opinions, motives, values, perceptions, cognitions, knowledge, psychology, 142, 363, 499, 522, 606, 628, 761, 809, 916, 965, 966, 969, 980, 996, 997, 1051, 1052, 1053, 1067, 1137, 1265, 1299, 1331, 1411, 1433, 1696, 1709, 1762, 1816, 1832, 1854, 1924, 1927, 1939, 1978, 1979, 2016, 2017, 2023, 2108, 2110

Automobile, 139, 140, 324, 383, 384, 776, 799, 803, 820, 1059, 1472, 1539, 1610, 1629, 1661, 1767, 1898, 1996, 2067, 2070, 2076, 2080, 2102

Built environment, urban environment, 65, 166, 367, 456, 554, 577, 578, 628, 766, 803, 932, 979, 1014, 1056, 1094, 1227, 1400, 1401, 1402, 1438, 1440, 1646, 1676, 1700, 1701, 1702, 1718, 1750, 1782, 1872, 1908, 1944, 1962, 1975, 2099

Canada, 7, 110, 145, 161, 218, 234, 316, 323, 373, 567, 592, 593, 842, 1018, 1178, 1194, 1238, 1353, 1454, 1791, 1795, 1810, 1811, 1812,

Canada--continued, 1822, 1843, 1855, 1864, 1867, 1869, 1871, 1872, 1875,
    1885, 1886, 1891, 1900, 1904, 1917, 1926, 1930, 1938, 1939, 1940,
    1946, 1973, 1982, 1983, 1985, 1993, 1996, 1998, 2009, 2013, 2017,
    2021, 2023, 2032, 2036, 2105, 2106, 2111, 2122, 2123

Coal/shale, 63, 78, 91, 122, 142, 160, 206, 212, 238, 239, 268, 300, 320,
    329, 346, 509, 550, 559, 560, 562, 590, 609, 632, 716, 723, 724,
    778, 918, 967, 1013, 1015, 1047, 1083, 1123, 1124, 1125, 1157, 1171,
    1181, 1190, 1198, 1234, 1235, 1266, 1285, 1360, 1394, 1471, 1480,
    1492, 1495, 1498, 1567, 1584, 1606, 1608, 1648, 1653, 1654, 1655,
    1708, 1740, 1760, 1785, 1792, 1794, 1803, 1809, 1819, 1820, 1837,
    1838, 1850, 1878, 1884, 1890, 1901, 1902, 1912, 1916, 1919, 1920,
    1921, 1927, 1931, 1932, 1933, 1942, 1948, 1956, 1961, 1963, 1966,
    1967, 1968, 1988, 1991, 2000, 2008, 2010, 2020, 2035, 2041, 2052,
    2068, 2094, 2104, 2107, 2114, 2118, 2124

Communication, media, 159, 539, 639, 776, 817, 944, 1150, 1254, 1256,
    1257, 1258, 1323, 1384, 1581, 1671, 1816

Conflicts, controversy, competition, 1, 6, 14, 20, 21, 22, 52, 58, 72,
    104, 111, 114, 136, 141, 148, 149, 164, 205, 228, 229, 235, 237,
    247, 274, 277, 283, 356, 368, 376, 409, 492, 494, 536, 538, 540,
    543, 549, 557, 563, 589, 590, 638, 641, 674, 680, 685, 689, 690,
    723, 724, 727, 737, 738, 751, 784, 814, 823, 844, 852, 875, 895,
    899, 966, 980, 988, 1002, 1042, 1050, 1051, 1052, 1053, 1054, 1058,
    1062, 1063, 1064, 1076, 1078, 1084, 1085, 1087, 1209, 1259, 1267,
    1316, 1372, 1375, 1383, 1396, 1408, 1412, 1449, 1464, 1466, 1467,
    1515, 1554, 1561, 1562, 1568, 1583, 1594, 1596, 1600, 1612, 1633,
    1669, 1695, 1744, 1764, 1785, 1789, 1808, 1817, 1827, 1853, 1857,
    1877, 1879, 1880, 1920, 1936, 1951, 1957, 1977, 1994, 2000, 2014,
    2016, 2038

Conservation, 40, 124, 125, 126, 127, 139, 140, 182, 255, 264, 367, 417,
    427, 456, 457, 493, 523, 551, 577, 578, 596, 597, 599, 600, 601,
    605, 607, 628, 643, 645, 647, 659, 690, 694, 695, 762, 766, 788,
    792, 819, 830, 833, 834, 859, 883, 887, 891, 924, 952, 962, 1044,
    1056, 1072, 1086, 1091, 1094, 1152, 1154, 1159, 1173, 1234, 1313,
    1400, 1401, 1402, 1431, 1442, 1500, 1508, 1517, 1536, 1569, 1616,
    1618, 1642, 1645, 1646, 1661, 1676, 1680, 1701, 1702, 1724, 1782,
    1862, 1882, 1894, 1901, 1908, 1928, 1964, 1980, 1991, 2092, 2099

Economics, business, industry, work, occupations, 10, 11, 12, 13, 14,
    15, 16, 25, 39, 49, 53, 55, 60, 66, 67, 72, 81, 84, 86, 99, 112,
    126, 132, 133, 139, 152, 156, 158, 162, 163, 167, 168, 172, 175,
    180, 181, 182, 184, 185, 186, 187, 188, 201, 203, 217, 220, 225,
    242, 249, 251, 252, 254, 258, 260, 262, 263, 290, 298, 299, 302,
    318, 330, 331, 332, 338, 339, 346, 354, 368, 369, 378, 382, 386,
    421, 430, 436, 445, 449, 452, 453, 454, 456, 459, 460, 462, 474,
    478, 482, 483, 486, 489, 491, 512, 514, 517, 518, 519, 521, 527,
    528, 529, 531, 532, 541, 544, 547, 550, 551, 559, 573, 583, 587,

Economics, etc.--continued, 589, 605, 608, 617, 619, 621, 624, 629, 630, 633, 635, 653, 670, 679, 690, 700, 703, 708, 710, 712, 716, 719, 725, 729, 730, 731, 746, 755, 758, 760, 771, 773, 778, 806, 807, 811, 825, 829, 831, 833, 840, 845, 850, 856, 859, 874, 889, 895, 903, 906, 907, 911, 912, 914, 918, 928, 929, 930, 937, 952, 957, 959, 964, 983, 991, 994, 1001, 1016, 1055, 1056, 1058, 1059, 1060, 1064, 1065, 1075, 1077, 1080, 1082, 1083, 1089, 1102, 1104, 1105, 1106, 1117, 1124, 1125, 1132, 1133, 1135, 1136, 1139, 1148, 1150, 1162, 1171, 1177, 1180, 1181, 1183, 1216, 1220, 1223, 1225, 1230, 1234, 1235, 1240, 1243, 1265, 1266, 1279, 1280, 1290, 1293, 1296, 1297, 1305, 1315, 1317, 1319, 1323, 1329, 1333, 1352, 1360, 1363, 1364, 1369, 1379, 1380, 1383, 1395, 1396, 1406, 1409, 1410, 1412, 1416, 1423, 1424, 1429, 1442, 1443, 1449, 1459, 1492, 1498, 1513, 1521, 1535, 1537, 1542, 1543, 1544, 1554, 1568, 1572, 1573, 1583, 1598, 1608, 1635, 1638, 1642, 1644, 1669, 1674, 1677, 1688, 1691, 1693, 1694, 1695, 1699, 1727, 1731, 1742, 1749, 1750, 1755, 1757, 1760, 1761, 1763, 1767, 1777, 1780, 1783, 1789, 1790, 1794, 1795, 1797, 1800, 1805, 1813, 1819, 1825, 1828, 1835, 1839, 1841, 1848, 1853, 1854, 1856, 1860, 1879, 1880, 1882, 1883, 1884, 1891, 1894, 1899, 1902, 1903, 1904, 1905, 1911, 1912, 1915, 1916, 1919, 1926, 1930, 1934, 1937, 1942, 1946, 1948, 1955, 1956, 1970, 1971, 1983, 2007, 2008, 2009, 2011, 2018, 2023, 2031, 2032, 2035, 2036, 2044, 2047, 2085, 2094, 2098, 2105, 2111, 2113, 2116, 2117, 2122, 2123

Education, 409, 442, 719, 802, 922, 924, 1144, 1256, 1384, 1671, 1778, 2017

Electricity, 4, 7, 9, 20, 27, 28, 29, 50, 51, 53, 54, 55, 60, 75, 76, 80, 86, 119, 132, 133, 168, 172, 188, 191, 210, 223, 233, 246, 247, 258, 288, 294, 299, 325, 354, 385, 387, 400, 405, 416, 443, 454, 554, 555, 558, 562, 583, 628, 639, 646, 648, 649, 650, 655, 669, 670, 684, 685, 689, 747, 764, 793, 794, 804, 847, 853, 862, 865, 895, 910, 932, 934, 935, 971, 1004, 1005, 1017, 1037, 1039, 1069, 1094, 1109, 1110, 1127, 1132, 1136, 1144, 1150, 1166, 1167, 1203, 1216, 1253, 1265, 1286, 1305, 1347, 1349, 1355, 1370, 1381, 1391, 1410, 1423, 1424, 1475, 1505, 1531, 1538, 1563, 1602, 1633, 1637, 1649, 1656, 1657, 1665, 1678, 1679, 1684, 1692, 1707, 1715, 1744, 1750, 1759, 1765, 1788, 1790, 1794, 1820, 1830, 1846, 1854, 1877, 1881, 1914, 1958, 1959, 1960, 1962, 1975, 1990, 2035, 2038, 2042, 2095

Energy crisis, 6, 11, 17, 24, 31, 34, 61, 72, 73, 86, 103, 148, 149, 150, 157, 163, 170, 171, 192, 193, 196, 197, 200, 211, 232, 235, 246, 256, 274, 275, 278, 279, 282, 283, 284, 289, 290, 300, 313, 321, 327, 337, 353, 358, 390, 391, 392, 396, 420, 439, 444, 455, 456, 463, 471, 472, 473, 474, 475, 494, 495, 496, 497, 502, 503, 516, 524, 530, 543, 565, 589, 609, 610, 627, 636, 659, 675, 688, 696, 698, 702, 718, 726, 740, 742, 744, 747, 761, 767, 774, 775, 780, 783, 785, 791, 796, 815, 820, 838, 839, 851, 852, 861, 863, 866, 871, 872, 877, 878, 879, 909, 926, 931, 936, 937, 943, 996,

Energy crisis--continued, 997, 1015, 1039, 1045, 1061, 1099, 1121, 1123,
1134, 1145, 1156, 1168, 1175, 1185, 1196, 1228, 1231, 1232, 1237,
1242, 1244, 1245, 1261, 1265, 1270, 1271, 1272, 1274, 1280, 1283,
1299, 1301, 1323, 1325, 1332, 1346, 1354, 1356, 1358, 1371, 1390,
1392, 1397, 1400, 1433, 1435, 1463, 1473, 1506, 1513, 1520, 1521,
1522, 1524, 1526, 1533, 1534, 1537, 1570, 1571, 1573, 1574, 1577,
1578, 1610, 1615, 1616, 1617, 1621, 1624, 1630, 1658, 1696, 1701,
1717, 1723, 1730, 1732, 1736, 1746, 1747, 1776, 1796, 1797, 1799,
1816, 1823, 1831, 1857, 1863, 1866, 1906, 1978, 1992, 2027, 2029,
2039, 2047, 2050, 2053, 2054, 2055, 2056, 2083, 2085, 2110

Energy, general, 5, 28, 33, 36, 45, 48, 50, 51, 56, 62, 74, 77, 79, 81,
88, 89, 101, 109, 117, 118, 123, 155, 164, 207, 213, 244, 248,
249, 253, 270, 272, 291, 322, 333, 334, 335, 336, 342, 345, 350,
372, 389, 403, 415, 435, 455, 458, 464, 467, 470, 488, 492, 501,
505, 506, 520, 525, 527, 528, 529, 582, 591, 602, 606, 613, 623,
629, 631, 647, 658, 660, 661, 664, 671, 672, 673, 682, 720, 743,
769, 773, 782, 802, 812, 841, 849, 855, 867, 869, 882, 944, 945,
970, 973, 1000, 1009, 1030, 1033, 1036, 1066, 1080, 1093, 1102,
1104, 1105, 1106, 1107, 1115, 1137, 1138, 1140, 1172, 1182, 1188,
1193, 1206, 1207, 1210, 1211, 1214, 1222, 1273, 1294, 1303, 1304,
1310, 1314, 1337, 1365, 1368, 1379, 1382, 1384, 1386, 1417, 1421,
1422, 1425, 1434, 1444, 1456, 1461, 1470, 1476, 1493, 1496, 1504,
1518, 1527, 1530, 1557, 1593, 1630, 1631, 1650, 1651, 1652, 1666,
1673, 1685, 1705, 1719, 1720, 1726, 1737, 1738, 1741, 1748, 1751,
1756, 1775, 1777, 1780, 1842, 1847, 1892, 1896, 1922, 1943, 1984,
1997, 2003, 2033, 2051, 2064, 2097, 2101, 2121

Energy industry associations, 31, 32, 33, 34, 35, 36, 37, 38, 39, 40,
41, 42, 43, 44, 46, 47, 48, 49, 50, 51, 73, 405, 406, 689, 690,
694, 695, 879, 1015, 1017, 1022, 1023, 1024, 1025, 1026, 1027,
1028, 1029, 1030, 1031, 1032, 1033, 1034, 1035, 1036, 1161, 1162,
1163, 1682, 1969, 1971

England, 9, 90, 91, 97, 175, 201, 292, 521, 534, 569, 587, 668, 707,
747, 938, 1003, 1047, 1073, 1165, 1350, 1452, 1706

Environment and energy, 3, 4, 5, 6, 20, 24, 29, 31, 35, 49, 58, 59, 75,
80, 94, 98, 99, 129, 131, 143, 145, 146, 150, 178, 183, 191, 194,
208, 210, 221, 223, 226, 233, 263, 267, 276, 319, 324, 344, 356, 359,
373, 374, 386, 388, 394, 398, 401, 406, 410, 425, 448, 452, 465,
480, 481, 490, 499, 508, 532, 533, 548, 557, 611, 612, 613, 618,
623, 629, 639, 641, 642, 646, 680, 689, 692, 693, 720, 725, 739,
745, 749, 751, 772, 791, 816, 851, 852, 853, 857, 880, 884, 885,
889, 897, 898, 902, 939, 941, 946, 947, 948, 950, 951, 953, 954,
955, 956, 961, 963, 977, 979, 981, 989, 993, 1005, 1006, 1008,
1028, 1040, 1062, 1063, 1067, 1069, 1070, 1073, 1078, 1079, 1086,
1104, 1105, 1106, 1109, 1111, 1128, 1129, 1144, 1155, 1166, 1172,
1185, 1191, 1195, 1201, 1203, 1208, 1215, 1218, 1223, 1225, 1226,
1237, 1256, 1257, 1258, 1259, 1268, 1277, 1281, 1286, 1288, 1293,

Environment and energy--continued, 1309, 1317, 1344, 1345, 1355, 1362,
  1376, 1377, 1430, 1436, 1439, 1445, 1450, 1452, 1462, 1475, 1505,
  1507, 1510, 1511, 1528, 1531, 1535, 1545, 1548, 1553, 1586, 1598,
  1603, 1636, 1638, 1647, 1649, 1653, 1662, 1674, 1677, 1699, 1700,
  1703, 1718, 1725, 1731, 1735, 1744, 1754, 1758, 1769, 1770, 1774,
  1778, 1779, 1785, 1786, 1796, 1810, 1815, 1824, 1825, 1826, 1829,
  1843, 1846, 1851, 1857, 1858, 1864, 1871, 1872, 1873, 1877, 1879,
  1881, 1897, 1898, 1903, 1906, 1916, 1929, 1976, 1977, 1986, 1987,
  1988, 2001, 2020, 2024, 2037, 2038, 2039, 2042, 2049, 2067, 2076,
  2080, 2086, 2093, 2094, 2095, 2109, 2110, 2112, 2116, 2120, 2124

Equity problems, 54, 132, 133, 242, 262, 320, 562, 624, 681, 685, 711,
  850, 865, 905, 907, 949, 1067, 1265, 1299, 1492, 1513, 1521, 1826,
  2021, 2023, 2047, 2080, 2085, 2108

Europe, Eastern (including Russia), 186, 199, 217, 256, 357, 381, 382,
  555, 580, 585, 614, 654, 674, 1180, 1189, 1280, 1716, 1775

Europe, Western, 60, 102, 177, 195, 196, 199, 241, 298, 423, 449, 559,
  560, 585, 763, 771, 836, 938, 1098, 1101, 1116, 1117, 1179, 1199,
  1278, 1361, 1410, 1424, 1976

Forests, 300, 322, 326, 329, 451, 1806, 1817, 1831, 2072

Fossil fuels, general, 10, 48, 158, 176, 294, 510, 569, 622, 638, 676,
  876, 930, 990, 1038, 1075, 1144, 1155, 1236, 1243, 1359, 1383,
  1448, 1494, 1512, 1540, 1541, 1547, 1550, 1555, 1587, 1593, 1630,
  1631, 1721, 1880, 1895, 2034, 2051, 2073

General ecology, social ecology, human ecology, eco-systems, 8, 94, 99,
  130, 146, 243, 272, 290, 318, 326, 363, 395, 518, 527, 528, 594,
  605, 611, 613, 623, 629, 657, 661, 705, 743, 772, 805, 882, 885,
  959, 989, 1104, 1106, 1137, 1148, 1172, 1210, 1211, 1259, 1268,
  1275, 1279, 1287, 1336, 1337, 1417, 1444, 1461, 1640, 1719, 1725,
  1726, 1737, 1738, 1754, 1786, 1815, 1872, 1973, 1989, 2006, 2037,
  2105, 2120

Geography, regional studies, 95, 97, 152, 160, 331, 357, 437, 526, 555,
  584, 637, 707, 827, 840, 896, 899, 904, 919, 1097, 1100, 1102,
  1158, 1192, 1266, 1278, 1335, 1453, 1552, 1711, 1812, 1813, 1819,
  1833, 1834, 1844, 1855, 1875, 1889, 1904, 1912, 1943, 1954, 1991,
  2032

Geothermal, 67, 68, 351, 429, 436, 1479, 1733

Government, public agencies, 13, 18, 26, 30, 43, 59, 85, 87, 105, 106,
  107, 111, 113, 114, 115, 116, 119, 122, 134, 141, 144, 165, 167,
  169, 170, 178, 179, 183, 185, 186, 189, 220, 224, 228, 233, 240,
  266, 274, 275, 277, 292, 293, 307, 309, 311, 320, 321, 341, 344,
  348, 355, 362, 364, 366, 371, 377, 407, 422, 424, 432, 435, 477,

Government, public agencies--continued, 484, 487, 495, 498, 500, 503,
  522, 540, 542, 543, 556, 564, 565, 568, 569, 571, 592, 593, 598,
  620, 652, 686, 699, 713, 714, 728, 730, 737, 741, 745, 757, 765,
  794, 801, 804, 813, 818, 822, 843, 844, 846, 847, 864, 868, 873,
  900, 911, 915, 925, 935, 943, 971, 975, 976, 984, 987, 988, 991,
  992, 995, 1004, 1038, 1046, 1076, 1077, 1081, 1082, 1091, 1108,
  1132, 1141, 1151, 1163, 1167, 1169, 1187, 1196, 1200, 1204, 1219,
  1221, 1233, 1250, 1263, 1264, 1284, 1286, 1316, 1330, 1333, 1347,
  1351, 1352, 1367, 1372, 1376, 1394, 1407, 1408, 1432, 1435, 1439,
  1446, 1447, 1464, 1471, 1474, 1498, 1499, 1504, 1509, 1523, 1542,
  1543, 1544, 1547, 1549, 1556, 1560, 1565, 1578, 1582, 1588, 1593,
  1597, 1601, 1611, 1613, 1614, 1617, 1625, 1653, 1669, 1670, 1672,
  1681, 1683, 1690, 1691, 1698, 1709, 1728, 1729, 1747, 1749, 1758,
  1766, 1786, 1807, 1843, 1861, 1923, 1950, 1965, 1972, 1994, 1999,
  2041, 2043, 2045, 2046, 2055, 2056, 2058, 2059, 2060, 2065, 2077,
  2084, 2089, 2092, 2093, 2097

Growth and energy, 54, 112, 168, 183, 198, 231, 248, 251, 286, 308, 360,
  361, 459, 462, 519, 529, 533, 579, 601, 772, 777, 786, 903, 1131,
  1139, 1213, 1223, 1287, 1293, 1296, 1421, 1449, 1598, 1724, 1755,
  1836, 1839, 1860, 1879, 1915

History, 111, 114, 216, 338, 426, 430, 500, 505, 586, 658, 714, 732,
  760, 928, 934, 1047, 1076, 1081, 1136, 1176, 1255, 1294, 1348,
  1416, 1427, 1448, 1449, 1451, 1592, 1609, 1728, 1739, 1819, 1888,
  1889, 1997

Hydroelectric, 634, 1158, 1176, 1177, 1186, 1316, 1407, 1408, 1474,
  1480, 1501, 1634, 1664, 1764, 1929, 2052

Import/export, 13, 41, 66, 69, 202, 218, 250, 262, 280, 281, 305, 483,
  486, 504, 592, 593, 729, 748, 824, 842, 929, 1018, 1025, 1078,
  1163, 1178, 1197, 1318, 1353, 1399, 1499, 1509, 1514, 1515, 1576,
  1619, 1891, 2048, 2065

International, interstate, intergovernmental, interagency, 11, 12, 13,
  15, 16, 17, 18, 30, 44, 52, 66, 69, 79, 81, 82, 90, 102, 111, 136,
  138, 148, 149, 164, 175, 177, 195, 196, 199, 201, 234, 235, 236,
  237, 242, 245, 251, 252, 254, 257, 292, 302, 321, 323, 332, 336,
  413, 423, 432, 434, 438, 440, 444, 447, 486, 504, 506, 515, 526,
  559, 580, 582, 583, 590, 603, 608, 619, 620, 630, 666, 674, 675,
  693, 697, 698, 712, 732, 748, 750, 771, 783, 797, 808, 813, 835,
  836, 869, 874, 875, 905, 907, 908, 910, 919, 921, 927, 930, 934,
  935, 945, 953, 985, 989, 1011, 1046, 1095, 1097, 1098, 1102, 1103,
  1107, 1108, 1113, 1115, 1117, 1120, 1121, 1122, 1126, 1134, 1146,
  1164, 1170, 1178, 1179, 1238, 1244, 1255, 1260, 1271, 1295, 1328,
  1361, 1374, 1390, 1396, 1398, 1399, 1409, 1412, 1423, 1426, 1435,
  1453, 1457, 1464, 1468, 1476, 1477, 1478, 1480, 1580, 1586, 1627,
  1643, 1658, 1659, 1698, 1743, 1766, 1774, 1802, 1804, 1845, 1887,
  2050, 2082, 2098

Land, 88, 122, 228, 230, 238, 239, 300, 457, 550, 562, 707, 723, 724, 933, 1190, 1258, 1348, 1560, 1597, 1653, 1703, 1718, 1799, 1831, 1850, 1852, 1888, 1954, 1996, 2002, 2012, 2020, 2024, 2025, 2026, 2106

Law, property rights, 44, 59, 75, 85, 88, 94, 108, 113, 134, 143, 144, 178, 179, 189, 202, 207, 224, 232, 233, 274, 275, 277, 304, 311, 355, 364, 379, 393, 410, 414, 415, 432, 434, 443, 444, 506, 515, 531, 556, 564, 568, 571, 573, 586, 590, 609, 625, 636, 684, 685, 686, 717, 741, 748, 749, 768, 783, 794, 804, 817, 822, 837, 843, 858, 864, 873, 894, 915, 917, 925, 933, 934, 935, 942, 943, 969, 988, 989, 994, 1003, 1054, 1062, 1069, 1077, 1078, 1109, 1142, 1150, 1167, 1176, 1203, 1208, 1219, 1221, 1264, 1319, 1340, 1355, 1394, 1407, 1414, 1426, 1466, 1474, 1504, 1506, 1507, 1508, 1518, 1524, 1528, 1545, 1547, 1548, 1571, 1575, 1578, 1581, 1584, 1586, 1595, 1600, 1611, 1615, 1616, 1625, 1626, 1687, 1688, 1767, 1792, 1807, 1947, 1953, 1954, 1965, 1969, 1994, 1999, 2000, 2001, 2002, 2015, 2026, 2041, 2043, 2045, 2053, 2056, 2057, 2059, 2060, 2063, 2073, 2079, 2084, 2090, 2091, 2093, 2104, 2118

Management, policy, decision making, planning, development, 2, 11, 13, 26, 30, 33, 37, 39, 40, 41, 42, 43, 44, 50, 51, 54, 59, 74, 85, 87, 88, 89, 90, 94, 95, 107, 108, 109, 119, 120, 121, 134, 151, 155, 159, 161, 166, 167, 178, 180, 186, 188, 195, 199, 202, 206, 207, 208, 219, 224, 231, 233, 234, 240, 241, 250, 274, 275, 276, 286, 287, 295, 296, 297, 304, 306, 307, 308, 309, 310, 311, 321, 323, 330, 331, 339, 341, 344, 347, 348, 355, 362, 364, 369, 371, 377, 379, 393, 397, 403, 405, 412, 416, 419, 421, 422, 424, 435, 442, 443, 460, 461, 467, 471, 475, 477, 479, 480, 484, 493, 495, 498, 503, 511, 515, 518, 535, 542, 545, 546, 547, 560, 567, 569, 576, 580, 591, 593, 598, 603, 604, 607, 620, 624, 627, 645, 652, 675, 678, 698, 699, 705, 706, 708, 709, 710, 713, 714, 728, 730, 737, 741, 745, 753, 764, 782, 795, 818, 823, 833, 834, 838, 848, 860, 861, 865, 868, 873, 875, 878, 886, 890, 900, 906, 920, 933, 939, 943, 946, 958, 972, 984, 991, 995, 1005, 1006, 1007, 1019, 1020, 1021, 1035, 1055, 1056, 1061, 1070, 1074, 1083, 1098, 1105, 1112, 1113, 1116, 1118, 1122, 1126, 1138, 1141, 1153, 1167, 1174, 1178, 1180, 1181, 1183, 1186, 1199, 1204, 1208, 1225, 1256, 1262, 1282, 1291, 1314, 1322, 1334, 1340, 1343, 1358, 1360, 1361, 1371, 1375, 1376, 1385, 1389, 1435, 1455, 1457, 1468, 1471, 1498, 1512, 1541, 1553, 1555, 1558, 1559, 1560, 1561, 1567, 1573, 1575, 1580, 1586, 1587, 1590, 1591, 1592, 1595, 1599, 1612, 1617, 1618, 1619, 1620, 1623, 1628, 1631, 1632, 1633, 1645, 1648, 1653, 1658, 1686, 1693, 1698, 1706, 1713, 1717, 1734, 1748, 1752, 1758, 1766, 1786, 1791, 1801, 1804, 1811, 1830, 1837, 1867, 1871, 1891, 1913, 1945, 1949, 1950, 1952, 1972, 1974, 2002, 2004, 2062, 2070, 2073, 2077, 2079, 2085, 2088, 2090, 2092, 2097, 2100, 2122

Mass/public transit, 23, 803, 964, 1411, 1440, 2055, 2070, 2076

Methodology, evaluation, measurement, indicators, systems analysis,
cost-benefits techniques, projections, monitoring, control,
standards, performance criteria, theory, concepts, research and
development, 7, 43, 46, 47, 53, 55, 62, 77, 105, 115, 120, 123,
124, 139, 140, 147, 154, 173, 182, 201, 214, 218, 225, 249, 250,
251, 253, 285, 288, 291, 294, 295, 296, 297, 299, 301, 326, 350,
359, 372, 378, 381, 387, 405, 419, 423, 449, 450, 452, 454, 460,
461, 505, 507, 512, 541, 544, 552, 577, 578, 579, 593, 597, 598,
601, 603, 605, 606, 622, 626, 631, 633, 635, 648, 670, 683, 715,
720, 722, 731, 735, 736, 746, 756, 757, 764, 770, 773, 777, 778,
779, 781, 785, 789, 818, 831, 832, 854, 886, 889, 906, 922, 950,
957, 958, 960, 961, 962, 963, 964, 968, 972, 982, 996, 997, 1000,
1012, 1017, 1018, 1036, 1041, 1091, 1093, 1119, 1139, 1140, 1187,
1188, 1193, 1217, 1224, 1225, 1236, 1241, 1275, 1283, 1287, 1292,
1294, 1318, 1322, 1344, 1358, 1359, 1371, 1387, 1388, 1391, 1413,
1458, 1476, 1482, 1483, 1484, 1487, 1489, 1495, 1496, 1497, 1516,
1519, 1525, 1549, 1559, 1575, 1577, 1585, 1605, 1607, 1649, 1652,
1659, 1660, 1663, 1671, 1681, 1707, 1734, 1756, 1780, 1783, 1784,
1789, 1794, 1803, 1814, 1821, 1830, 1835, 1841, 1847, 1859, 1883,
1887, 1891, 1897, 1898, 1910, 1911, 1913, 1949, 1959, 1960, 1962,
1970, 1983, 1984, 1990, 2035, 2061, 2078, 2079, 2094, 2121

Middle East/Arabs, 11, 30, 69, 82, 135, 136, 137, 222, 245, 321, 380,
413, 447, 472, 482, 580, 587, 620, 666, 674, 701, 703, 783, 787,
808, 809, 836, 907, 919, 926, 927, 928, 929, 985, 1049, 1231,
1255, 1295, 1328, 1412, 1627, 1657, 1694, 1698

Military, war, security, 10, 116, 312, 348, 349, 353, 400, 465, 500,
615, 787, 800, 801, 812, 830, 913, 1049, 1163, 1227, 1365, 1377,
1399, 1441, 1469, 1499, 1509, 1579, 1656, 1672, 1685, 1690, 1713,
1743, 1745, 1804, 1808, 1865, 2030, 2050

Natural or bottle gas, 32, 33, 46, 84, 96, 134, 170, 185, 217, 229,
257, 266, 269, 307, 352, 355, 411, 476, 477, 484, 485, 491, 507,
516, 520, 556, 561, 566, 592, 593, 626, 630, 650, 686, 690, 694,
711, 717, 746, 752, 781, 806, 821, 832, 842, 844, 848, 858, 860,
873, 894, 915, 934, 942, 1023, 1028, 1029, 1034, 1044, 1048, 1065,
1071, 1096, 1098, 1117, 1127, 1161, 1169, 1216, 1219, 1249, 1250,
1263, 1347, 1353, 1397, 1406, 1427, 1460, 1464, 1480, 1497, 1551,
1561, 1566, 1576, 1612, 1619, 1620, 1667, 1668, 1669, 1677, 1687,
1695, 1710, 1729, 1753, 1787, 1791, 1807, 1810, 1812, 1834, 1855,
1861, 1870, 1891, 1902, 1912, 1923, 1946, 2009, 2021, 2046, 2057,
2063, 2074

Natural resources, general, 42, 44, 112, 147, 176, 212, 263, 301, 339,
379, 397, 451, 457, 511, 624, 635, 637, 676, 677, 715, 719, 731,
757, 773, 777, 779, 797, 781, 838, 864, 868, 876, 900, 951, 954,
956, 980, 993, 1010, 1021, 1048, 1108, 1129, 1130, 1135, 1139,
1366, 1372, 1375, 1413, 1455, 1508, 1509, 1510, 1512, 1540, 1541,
1546, 1565, 1582, 1587, 1598, 1618, 1632, 1644, 1663, 1672, 1675,
1727, 1749, 1798, 1809, 1813, 1835, 1874, 1895, 1897, 1910, 1915,
1937, 1947, 1952, 2002, 2018, 2026, 2034, 2051

North Sea, 97, 301, 433, 434, 521, 771, 938

Nuclear, 25, 35, 36, 70, 71, 98, 104, 121, 151, 158, 161, 165, 178, 181, 189, 190, 201, 205, 221, 237, 240, 241, 271, 285, 328, 341, 349, 371, 376, 390, 391, 392, 398, 399, 402, 410, 438, 465, 466, 468, 469, 530, 536, 537, 538, 539, 540, 552, 559, 570, 571, 572, 573, 574, 581, 621, 637, 651, 656, 662, 687, 691, 692, 721, 722, 733, 745, 754, 770, 784, 789, 800, 801, 814, 815, 816, 817, 828, 845, 862, 898, 902, 904, 916, 923, 925, 941, 955, 960, 975, 978, 987, 988, 991, 992, 1002, 1042, 1050, 1052, 1053, 1054, 1058, 1073, 1074, 1084, 1085, 1087, 1088, 1089, 1090, 1114, 1120, 1143, 1144, 1179, 1184, 1189, 1200, 1202, 1209, 1213, 1214, 1217, 1227, 1233, 1239, 1240, 1251, 1254, 1258, 1264, 1269, 1289, 1290, 1297, 1306, 1307, 1308, 1311, 1320, 1342, 1357, 1362, 1369, 1385, 1419, 1420, 1437, 1441, 1451, 1466, 1467, 1469, 1480, 1481, 1482, 1483, 1484, 1485, 1486, 1487, 1488, 1489, 1490, 1491, 1502, 1503, 1529, 1532, 1600, 1693, 1722, 1743, 1745, 1825, 1864, 1865, 1869, 1877, 1886, 1925, 1936, 1957, 1981, 2006, 2040, 2043, 2044

Offshore/ocean sources, 44, 97, 134, 180, 184, 216, 280, 304, 308, 365, 476, 665, 734, 735, 901, 965, 966, 968, 1024, 1031, 1319, 1324, 1403, 1445, 1464, 1562, 1687, 1768, 1771, 1773, 1909, 2069, 2082

Oilspills, 110, 216, 230, 308, 359, 404, 531, 534, 665, 668, 901, 912, 917, 965, 966, 968, 1003, 1165, 1326, 1350, 1403, 1426, 1428, 1459, 1662, 1687, 1768, 1771, 1773, 1829, 1909, 1917, 1940, 2069, 2082

Petroleum, 11, 12, 13, 15, 16, 17, 19, 23, 37, 38, 39, 40, 41, 42, 43, 44, 46, 47, 49, 52, 66, 69, 82, 83, 94, 97, 110, 111, 113, 114, 128, 134, 135, 136, 137, 138, 143, 162, 166, 175, 177, 184, 186, 187, 202, 203, 204, 217, 219, 222, 230, 242, 245, 250, 251, 252, 254, 260, 262, 265, 266, 274, 275, 280, 281, 301, 302, 305, 310, 311, 312, 339, 347, 353, 369, 370, 379, 380, 381, 382, 384, 413, 414, 423, 424, 426, 430, 431, 432, 433, 434, 437, 440, 445, 447, 453, 472, 473, 476, 482, 485, 486, 488, 500, 504, 511, 521, 531, 561, 566, 580, 587, 590, 615, 619, 630, 649, 657, 665, 666, 668, 674, 680, 694, 695, 697, 699, 701, 702, 703, 706, 709, 714, 729, 732, 741, 748, 749, 750, 752, 755, 758, 759, 760, 765, 771, 776, 778, 787, 790, 798, 799, 808, 813, 820, 824, 829, 830, 831, 832, 833, 834, 836, 842, 843, 857, 858, 859, 860, 874, 879, 892, 893, 894, 901, 905, 907, 911, 913, 914, 919, 928, 929, 930, 933, 938, 942, 943, 965, 966, 968, 974, 985, 986, 1022, 1023, 1024, 1025, 1026, 1027, 1028, 1029, 1030, 1031, 1034, 1035, 1044, 1056, 1076, 1078, 1081, 1095, 1097, 1098, 1100, 1101, 1119, 1145, 1146, 1147, 1160, 1161, 1162, 1163, 1164, 1170, 1174, 1178, 1194, 1197, 1205, 1215, 1228, 1229, 1230, 1248, 1250, 1255, 1262, 1266, 1267, 1268, 1291, 1295, 1315, 1318, 1326, 1328, 1329, 1335, 1341, 1346, 1363, 1364, 1365, 1395, 1396, 1399, 1412, 1426, 1427, 1428, 1447, 1464, 1468, 1472, 1480, 1499, 1506, 1509, 1514, 1515, 1537, 1542, 1543,

Petroleum--continued, 1544, 1552, 1566, 1576, 1577, 1579, 1583, 1585,
1613, 1619, 1620, 1622, 1627, 1628, 1629, 1635, 1641, 1643, 1648,
1653, 1657, 1669, 1682, 1686, 1694, 1698, 1711, 1712, 1725, 1728,
1742, 1755, 1760, 1786, 1791, 1802, 1829, 1834, 1909, 1912, 1917,
1930, 1935, 1940, 1948, 1950, 1956, 1971, 2047, 2048, 2057, 2058,
2065, 2066, 2068, 2069, 2082, 2083, 2087, 2098

Politics, political science, 8, 11, 15, 18, 19, 27, 61, 72, 99, 101,
114, 148, 149, 153, 156, 165, 180, 195, 196, 199, 201, 204, 208,
222, 245, 259, 266, 309, 315, 370, 424, 433, 434, 435, 440, 475,
497, 504, 546, 561, 587, 589, 619, 625, 666, 693, 701, 732, 742,
750, 752, 759, 771, 795, 796, 797, 800, 808, 838, 874, 901, 908,
929, 983, 985, 994, 1046, 1052, 1053, 1076, 1077, 1097, 1104,
1135, 1146, 1228, 1229, 1248, 1288, 1328, 1373, 1374, 1375, 1411,
1412, 1455, 1468, 1793, 2007

Pollution, general, 35, 83, 129, 131, 145, 210, 221, 273, 346, 347,
359, 383, 397, 435, 460, 512, 553, 638, 642, 685, 768, 810, 881,
993, 999, 1062, 1111, 1138, 1142, 1155, 1166, 1201, 1305, 1638,
1718, 2055

Population, demography, migration, crowding, 333, 451, 687, 1220, 1287,
1510, 1598, 1700, 1860, 1889, 1938, 1985, 1991, 2023, 2037

Quality of life, affluence, living standards, general environmental
quality, life-style, 65, 87, 155, 198, 303, 401, 418, 442, 529,
870, 888, 967, 1012, 1128, 1203, 1269, 1298, 1798, 1818, 1827,
1878, 1890, 1924, 1927, 1931, 1932, 1933, 1974, 1978, 1979, 2006,
2027

Railroads, 100, 340, 459, 974, 1194, 1930

Readers, conference proceedings, special issues, textbooks, bibliographies, 19, 29, 36, 48, 63, 75, 79, 113, 117, 131, 154, 173, 176,
185, 192, 193, 198, 200, 216, 220, 221, 226, 270, 279, 284, 318,
342, 347, 351, 359, 373, 387, 394, 407, 408, 409, 417, 418, 428,
490, 520, 557, 591, 634, 665, 688, 691, 736, 758, 772, 788, 795,
815, 883, 944, 951, 954, 956, 981, 995, 1040, 1045, 1066, 1129,
1138, 1151, 1156, 1158, 1164, 1174, 1191, 1199, 1207, 1248, 1258,
1274, 1282, 1285, 1300, 1303, 1304, 1348, 1415, 1427, 1435, 1455,
1458, 1475, 1526, 1588, 1589, 1594, 1595, 1598, 1599, 1603, 1604,
1605, 1617, 1626, 1630, 1662, 1685, 1730, 1754, 1758, 1781, 1784,
1801, 1802, 1807, 1824, 1838, 1839, 1852, 1859, 1866, 1873, 1874,
1876, 1881, 1921, 1922, 1963, 1969, 1980, 1986, 1989, 2004, 2006,
2012, 2026, 2028, 2034, 2096, 2103, 2112

Reclamation, 144, 160, 212, 238, 239, 300, 329, 490, 550, 562, 723,
724, 1013, 1043, 1190, 1606, 1785, 1792, 1799, 1805, 1806, 1821,
1831, 1844, 1848, 1873, 1876, 1893, 1911, 1918, 1968, 1976, 1987,
2005, 2011, 2040, 2091, 2094, 2096, 2104

Recreation, leisure, parks, wilderness, wildlife, nature, 42, 398, 406, 641, 723, 798, 853, 870, 899, 912, 1003, 1043, 1165, 1215, 1350, 1366, 1408, 1459, 1768, 1771, 1773, 1906, 1917, 1940, 1996, 2005, 2019, 2020

Safety, health, 35, 104, 151, 165, 189, 190, 194, 221, 268, 271, 328, 349, 399, 441, 448, 465, 466, 536, 537, 538, 539, 540, 549, 552, 553, 571, 572, 573, 575, 662, 722, 733, 793, 828, 904, 925, 978, 1200, 1233, 1251, 1269, 1289, 1357, 1437, 1441, 1466, 1469, 1745, 1825, 1865, 1925, 1981, 2006, 2040, 2124

Science, technology, 56, 77, 105, 121, 125, 139, 154, 173, 212, 220, 224, 314, 354, 367, 376, 389, 397, 504, 513, 531, 570, 598, 663, 682, 716, 721, 735, 736, 756, 800, 957, 961, 1002, 1008, 1052, 1053, 1054, 1068, 1139, 1173, 1224, 1225, 1252, 1302, 1389, 1416, 1418, 1423, 1559, 1655, 1676, 1697, 1732, 1739, 1784, 1793, 1809, 1832, 1898, 1948, 1970, 1971, 2035, 2064, 2115, 2119

Siting, 20, 21, 141, 152, 178, 179, 210, 228, 364, 374, 388, 562, 568, 618, 684, 721, 727, 738, 794, 828, 837, 992, 1051, 1052, 1053, 1054, 1218, 1247, 1338, 1340, 1355, 1362, 1414, 1439, 1528, 1704, 1715, 1744, 1769, 1770, 1788, 1854, 1990, 2049

Sociology, social organization, institutions, culture, society, 8, 19, 20, 21, 22, 119, 157, 206, 209, 212, 226, 238, 239, 243, 270, 290, 295, 303, 318, 362, 395, 417, 427, 446, 480, 499, 508, 519, 550, 577, 578, 589, 594, 616, 628, 629, 661, 670, 676, 715, 716, 736, 740, 754, 792, 805, 807, 875, 883, 906, 910, 918, 965, 966, 967, 980, 993, 996, 997, 1012, 1055, 1083, 1104, 1114, 1115, 1124, 1125, 1143, 1157, 1162, 1177, 1210, 1211, 1220, 1248, 1253, 1281, 1298, 1299, 1331, 1360, 1362, 1370, 1372, 1373, 1374, 1377, 1404, 1411, 1439, 1461, 1467, 1492, 1662, 1696, 1715, 1722, 1726, 1737, 1738, 1739, 1760, 1762, 1798, 1809, 1812, 1815, 1816, 1832, 1854, 1875, 1878, 1889, 1903, 1905, 1924, 1927, 1931, 1932, 1933, 1939, 1942, 1955, 1973, 1974, 1982, 1983, 1985, 1991, 1998, 2007, 2009, 2021, 2023, 2027, 2032, 2105, 2115, 2123

Solar, 64, 69, 174, 441, 588, 653, 826, 827, 1011, 1041, 1212, 1398, 1479, 2061, 2119

Strip mining, 34, 122, 144, 160, 212, 238, 239, 300, 320, 329, 490, 499, 550, 562, 586, 723, 724, 983, 1013, 1043, 1055, 1083, 1124, 1125, 1157, 1198, 1378, 1495, 1512, 1564, 1594, 1606, 1647, 1648, 1653, 1654, 1780, 1785, 1792, 1798, 1799, 1803, 1805, 1806, 1809, 1817, 1818, 1827, 1831, 1844, 1848, 1850, 1873, 1876, 1878, 1884, 1888, 1889, 1890, 1893, 1903, 1905, 1911, 1916, 1918, 1920, 1921, 1924, 1931, 1932, 1933, 1934, 1941, 1942, 1947, 1949, 1953, 1954, 1955, 1965, 1969, 1976, 1987, 1988, 1989, 1991, 1994, 1999, 2000, 2005, 2010, 2011, 2014, 2015, 2019, 2020, 2024, 2025, 2034, 2059, 2060, 2072, 2073, 2075, 2084, 2091, 2094, 2096, 2104, 2107, 2112, 2114

Supply/demand/price, 11, 12, 13, 14, 15, 16, 18, 23, 27, 32, 39, 40, 45, 46, 53, 62, 63, 72, 84, 96, 112, 116, 118, 128, 129, 132, 133, 135, 137, 163, 170, 183, 186, 202, 213, 218, 246, 248, 251, 253, 257, 258, 262, 269, 280, 281, 282, 283, 288, 291, 294, 298, 299, 302, 307, 312, 319, 325, 334, 335, 336, 338, 352, 357, 360, 361, 372, 375, 380, 417, 423, 426, 430, 431, 437, 438, 445, 451, 453, 454, 458, 476, 477, 478, 482, 484, 491, 494, 495, 496, 516, 525, 526, 566, 577, 578, 579, 597, 601, 610, 612, 648, 651, 654, 656, 667, 688, 690, 697, 704, 705, 711, 725, 746, 752, 763, 769, 775, 779, 780, 781, 782, 785, 786, 789, 792, 821, 829, 830, 836, 839, 844, 848, 855, 860, 861, 862, 867, 869, 877, 878, 879, 883, 891, 892, 895, 913, 971, 973, 977, 1015, 1017, 1018, 1022, 1027, 1033, 1034, 1056, 1071, 1090, 1096, 1099, 1101, 1110, 1127, 1139, 1140, 1150, 1168, 1169, 1193, 1206, 1237, 1241, 1246, 1249, 1270, 1312, 1321, 1351, 1352, 1356, 1373, 1375, 1382, 1397, 1419, 1429, 1438, 1447, 1453, 1460, 1465, 1470, 1476, 1477, 1478, 1493, 1497, 1525, 1551, 1552, 1579, 1605, 1622, 1624, 1627, 1641, 1654, 1667, 1668, 1682, 1688, 1690, 1699, 1704, 1710, 1716, 1721, 1741, 1750, 1772, 1783, 1787, 1790, 1820, 1833, 1834, 1840, 1849, 1861, 1868, 1870, 1879, 1923, 1935, 1944, 1958, 1959, 1960, 1961, 1962, 1966, 1967, 1970, 1975, 1997, 2003, 2013, 2030, 2038, 2074, 2087, 2102

Tankers/tanker spills, 110, 230, 404, 534, 665, 668, 901, 1003, 1165, 1326, 1350, 1428, 1459, 1558, 1623, 1643, 1768, 1771, 1773, 1768, 1829, 2082

Taxation, 11, 37, 39, 167, 339, 369, 379, 431, 511, 729, 858, 864, 881, 942, 1161, 1162, 1250, 1263, 1498, 1620, 1800, 1801, 1874, 1895, 1910, 1919, 1937, 2010, 2107

Utilities, 7, 26, 27, 28, 29, 50, 51, 80, 84, 86, 98, 119, 141, 146, 153, 179, 183, 288, 289, 292, 293, 317, 319, 385, 405, 406, 487, 491, 517, 519, 522, 547, 548, 554, 564, 568, 583, 617, 618, 626, 639, 649, 650, 683, 689, 690, 705, 725, 727, 733, 739, 804, 821, 837, 846, 847, 881, 884, 915, 935, 971, 1004, 1017, 1039, 1111, 1127, 1132, 1187, 1191, 1204, 1216, 1218, 1226, 1247, 1286, 1305, 1322, 1330, 1334, 1349, 1370, 1391, 1393, 1410, 1446, 1475, 1503, 1505, 1538, 1549, 1563, 1586, 1625, 1656, 1665, 1666, 1678, 1679, 1685, 1691, 1692, 1695, 1699, 1759, 1914, 2071

Voluntary action, voluntary organizations, citizen participation, social movements, 20, 21, 22, 104, 216, 233, 264, 376, 406, 468, 536, 538, 591, 660, 721, 738, 739, 756, 784, 814, 851, 852, 953, 965, 966, 980, 988, 997, 1002, 1042, 1050, 1051, 1052, 1053, 1054, 1200, 1221, 1286, 1330, 1331, 1338, 1340, 1378, 1393, 1403, 1420, 1436, 1466, 1500, 1713, 1736, 1785, 1806, 1830, 1972, 1979, 2004, 2016, 2024

Waste, 92, 139, 349, 553, 766, 788, 816, 1014, 1072, 1262, 1459, 1688, 1911, 1925, 1928, 2011

Water, 70, 76, 110, 152, 216, 227, 299, 308, 346, 374, 378, 400, 404, 534, 553, 558, 581, 604, 665, 668, 684, 708, 716, 724, 741, 899, 901, 906, 912, 917, 965, 966, 1003, 1050, 1078, 1158, 1165, 1254, 1258, 1326, 1348, 1350, 1403, 1408, 1428, 1445, 1450, 1458, 1459, 1501, 1634, 1687, 1714, 1764, 1768, 1771, 1773, 1852, 1876, 1909, 1988, 1989, 2029, 2035, 2052

Western/Southwestern U.S., 20, 22, 54, 55, 95, 114, 132, 133, 142, 160, 211, 216, 229, 277, 300, 307, 312, 320, 330, 331, 361, 403, 442, 456, 499, 562, 719, 723, 724, 764, 794, 807, 899, 906, 912, 918, 965, 966, 967, 1013, 1043, 1055, 1076, 1083, 1124, 1125, 1181, 1247, 1258, 1360, 1366, 1403, 1411, 1492, 1498, 1563, 1602, 1648, 1727, 1740, 1764, 1788, 1792, 1827, 1830, 1831, 1837, 1838, 1850, 1854, 1876, 1878, 1890, 1920, 1931, 1932, 1933, 1943, 1958, 1959, 1960, 1966, 1968, 1974, 1975, 1987, 1988, 1989, 1990, 2025, 2035, 2069, 2072, 2074, 2096, 2107, 2112

Thermal energy, thermal pollution, 70, 76, 227, 514, 558, 581, 604, 684, 828, 1091, 1221, 1381, 1490, 1858, 2035

Transportation, general, 23, 215, 481, 563, 640, 643, 645, 891, 940, 947, 948, 950, 1092, 1276, 1494, 1749, 1862, 1964, 2036, 2067, 2088

Z
5853
P83
M62

AUG 11 1976